"What I found so enchanting about [the] book is that it's full of people like [her] . . . keepers of the planet."
–Bill Moyers

"Moving . . . For Goodall, both hubris and tragedy reside in the fact that extinctions are being allowed to happen. The question is: What to do about it? In her book, there's a take-home response."
—CSMonitor.com (*Christian Science Monitor*)

"A testament of hope, an inspirational call to transcend our history of ignorance and neglect and to move in the direction of communal good will on behalf of nature. It illustrates through vivid, heroic examples how humankind's fundamental impulse to heal has already saved dozens of species from the brink of extinction."
—*Austin American-Statesman*

"The good news is, to break the doom-and-gloom cycle of cynicism, we have Dr. Jane Goodall . . . to offer a remarkably optimistic point of view in her book . . . It's a 'you can do it' ecological pep talk."
—*Boston Phoenix*

"Goodall is no Pollyanna about species reclamation . . . but these accounts of conservation success are inspirational."
—*Publishers Weekly*

"Renowned scientist Jane Goodall brings us inspiring news . . . as well as fascinating survival stories."
—*More*

Hope
for Animals
and Their World

Hope
for Animals
and Their World

HOW ENDANGERED SPECIES ARE
BEING RESCUED FROM THE BRINK

Jane Goodall
with Thane Maynard
and Gail Hudson

GRAND CENTRAL
PUBLISHING

NEW YORK   BOSTON

Grand Central Publishing
Hachette Book Group
237 Park Avenue
New York, NY 10017

www.HachetteBookGroup.com

Printed in the United States of America

Originally published in hardcover by Grand Central Publishing.

First Trade Edition: June 2011
10 9 8 7 6 5 4 3 2 1

Grand Central Publishing is a division of Hachette Book Group, Inc.
The Grand Central Publishing name and logo is a trademark of
Hachette Book Group, Inc.

The publisher is not responsible for websites (or their content) that are not owned by
the publisher.

The Library of Congress has cataloged the hardcover edition as follows:

Goodall, Jane
    Hope for animals and their world : how endangered species are being rescued
from the brink / Jane Goodall with Thane Maynard and Gail Hudson.
        p.   cm.
        Includes bibliographical references.
        ISBN 978-0-446-58177-6 (regular edition)—ISBN 978-0-446-55994-2 (large
    print edition)   1. Endangered species.   2. Wildlife rescue.   3. Wildlife
    reintroduction.   4. Nature conservation.   I. Maynard, Thane.   II. Hudson,
    Gail E.   III. Title.

    QH75.G636
    591.68—dc22   2009011215

ISBN 978-0-446-58178-3 (pbk.)

Dr. Jane Goodall and the Institute do not endorse handling or interfering with wild
chimpanzees. The chimpanzee in the jacket photo has been orphaned and lives at a
sanctuary.

This book was printed on Domtar's EarthChoice Tradebook. The paper contains
fiber from well-managed environmentally sound forests that are independently
certified to the standards of the Forest Stewardship Council. This paper also contains
25% recycled paper, which was recovered from post-consumer waste.

This book is dedicated to the memory of Martha,
the last passenger pigeon—and to the last Miss Waldron's
colobus and the last Yangtze River dolphin. As we think of
their lonely end, may we be inspired to work harder
to prevent others suffering a similar fate.

Contents

Acknowledgments

This book was several years in the making, and it could not have been written without the help of many people. Indeed, one of the really great experiences for me during the past few years has been meeting so many extraordinary and dedicated scientists and conservationists. Between them they have accomplished so much, and I have been overwhelmed by their willingness to share their knowledge and to read, correct, and add to the accounts I had written about their projects. Such generosity. I cannot thank them enough.

While writing this book I learned about many wonderful projects from around the world. Unfortunately, when they were all written up, it was obvious that the manuscript was too long. Even when each story was cut, and cut again, the book was still too long. After much agonizing, it was decided that the only thing we could do was remove whole sections. I still feel devastated that this had to be done, mostly because the people whose projects I was writing about had spent so much time reading their chapters and ensuring that the information was correct—and they were so pleased that the material would be included in the book. I know they will be disappointed and I feel terrible about it.

Our Web Site: However, there is a silver lining. The publishers agreed to create a Web site which features all this material (janegoodall hopeforanimals.com). It also includes the original versions of some chapters that were shortened for the book, as well as many photographs we gathered. I encourage everyone to visit the Web site and learn about the wonderful projects described there. You will also find my complete acknowledgments that had to be condensed for this book.

As I said, without the help and ongoing cooperation of the people profiled in these pages, this book would never have been possible. You will find their names and heroic stories in the chapters to come. I would also like to thank the following people, who helped us tremendously but whose names you won't find in the pages ahead: Mark Bain (short-nosed sturgeon), Ann M. Burke (whooping crane), Phil Bishop (Hamilton frog), Pat Bowles (Caspian horse), Jane Chandler (whooping crane), Glenn Fraser (woodhen), Rod Gritten (glutinous snail), Nancy Haley (short-nosed sturgeon), Kirk Hart (short-tailed albatross), Diane Hendry (red wolf), Dave Jarvis (pedder galaxias), Tom Koerner and Dan Miller (trumpeter swan), Bill Lautenbach (Sudbury, Ontario), Alfonso Aguirre Muñoz (Guadalupe Island), Mark Stanley Price (Arabian oryx), Ken Reininger (nene), Ruth Shea (trumpeter swan), Amy Sprunger (moapa dace), John Thorbjarnarson (Chinese alligator), Mike Wallace (California condor), Jake Wickerham (pedder galaxias), and Stephen S. Young (Cao Hai Nature Reserve).

I am really grateful to Don Merton. He has helped us with so many chapters in this book: The utterly fascinating story of the kakapo, New Zealand's large flightless parrot, will appear on our Web site. And I thank Nicholas Carlile for his enormous help in reviewing several of the stories in this book. His contribution to rescuing the Gould's petrel will also appear on our Web site.

The following people provided me with information about heroic efforts to save our endangered plant species: Peter Raven, Hugh Bollinger, Nick Johnson, Lourdes "Lulu" Rico Arce, Michael Park, Tim Rich, Bill Brumback, Jo Meyerkord, Kathryn Kennedy, and Robin Wall Kimmerer. You will find their stories and contributions on our Web site. In particular, Victoria Wilman and Robert Robichaux sent me so much helpful information, as did Paul Scannell and Andrew Pritchard, who also met with me in Australia.

Another whole section that we couldn't fit into the book but will appear on our Web site is about how the general public and our youth are helping to save endangered species. It describes fabulous stories of how an endangered species can halt developments: Greg Ballmer told me about the Delhi Sands flower-loving fly, and I learned about the Salt Creek tiger beetle from Stephen Spomer, Leon Higley, Mitch Paine, and Jessa Huebing-Reitinger. Matt and Ann Magoffin are helping to save the Chirakua leopard frog, and Meredith Dreifus and her family are helping the red-cockaded woodpecker. Information for the Roots & Shoots section was provided by Chase Pickering, Tony Liu, and Dan Fulton. I was also helped by Susan and Alexandra Morris and Tim Coonan, who have worked for many years to protect the Channel Island fox.

Thane Maynard: I was extremely fortunate to meet with and interview a wide and fiercely bright cast of characters in the process of developing this book. Each of these scientists and conservationists has stood in the gap when it mattered most for these species. I would like to acknowledge the following people who helped me gather my Field Notes but whose stories and names do not appear in these pages. All of these people and stories will be featured on our Web site: Wangari Mathaai and her staff from the Greenbelt Movement; Kent Vliet (the American alligator); Pete Dunne (bald eagle); Rick McIntyre (gray wolf); Clay Degayner (Key Largo woodrat); Ron Austing (Kirtland's warbler); Scott Eckert (leatherback sea turtle); Greg Neudecker (trumpeter swan); Geoff Hill (ivory-billed woodpecker); Roger Payne (Pacific gray whale); Greg Sherley (weta of New Zealand); and Michael Samways (South African dragonflies). Naturally I want to thank my wife, Kathleen, for all her help and support through the years I worked on this book. My thanks go as well to the remarkable staff at the Cincinnati Zoo & Botanical Garden, who earn their stripes inspiring every zoo visitor with wildlife every day.

Gail Hudson: Many thanks to my agent, Mary Ann Naples of the Creative Culture, for her outstanding support and guidance. And I am especially grateful to my husband, Hal, daughter, Gabrielle, and son, Tennessee, who always support my work in the world.

Photography: All the photographs that you see in this book and on our Web site were donated to us by the photographers. We are deeply grateful to all of them for their generosity and support. You will find their names in the photo credits, alongside their photographs. In many cases the heroes we profiled assisted us in finding photographs, but we also want to thank and acknowledge the following people who helped us acquire photos: Shalese Murray, Andrew Bennet, JoGayle Howard, Gary Fry, Fr. Ed Udovic, C.M., James Popham, Ann Burke, Christina Anderson, Douglas W. Smith, Antonio Rivas, Christina Simmons, Caron Glover, Penny Haworth, Vanessa Dinning, Stephen Monet, Jesse Grantham, Liz Condie, David van Berkel, and Rob Robichaux.

JGI and Worldwide Helpers: Throughout the writing of this book, and our search for information and photographs, the following staff members from our different JGI offices around the world were extraordinarily helpful: Federico Bogdanowicz, Ferran Guallar, David Lefrance, Jeroen Haijtink, Polly Cevallos, Kelly Kok, Walter Inmann, Gudrun Schindler, Melissa Tauber, Claire Quarendon, Anthony Collins, Grace Gobbo, Jane Lawton, Sophie Muset, Erika Helms, Zhang Zh, Michael Crook, and Greg MacIsaac.

I wish there was space to thank every one of our JGI staff members on the TACARE restoration program around the Gombe National Park. But I must mention Emmanuel Mtiti, Mary Mavanza, Aristedes Kashula, and Amani Kingu, who helped us with the material that appears here and on our Web site.

In the early stages of the book JGI volunteer Joy Hotchkiss helped with research and preliminary interviews, and Sally Eddows developed products featuring endangered species that will help to promote the book. We are extremely grateful to Mary Paris, who edited all the photos that appear in the book and on our Web site. And Meredith Bailey, editorial assistant to Gail, helped us with the "What You Can Do" section, as did Claire Jones of JGI.

I am filled with gratitude to the staff of the Global Office of the Founder (GOOF). In particular, Rob Sassor contacted a great many people during the first few years of the book, interviewing them and providing me with information; he was really enthused by this project, and his help was invaluable. Stephen Ham, who stepped into Rob's position, also helped to contact scientists and organize meetings. Susana Name, who helps to manage my hectic schedule, somehow arranged to fit in meetings with scientists involved in rescuing some of the species discussed in this book.

There is absolutely no way that the photographs for this book could have been gathered from farflung corners of the globe, organized, and evaluated without the dedicated, detailed, and persistent efforts of Christin Jones. She never gave up on getting an image. She was terrific to work with. And she was indefatigable—even major surgery could not, for long, keep her from putting the last photos in order. What a hero!

Grand Central: We owe a huge debt of gratitude to the staff at Grand Central, people who were supportive and understanding throughout the years of making this book. Editor Natalie Kaire stayed in close contact with us and went through the original long version of the book a few times, helping us make tough decisions about what should be cut, from both text and photographs. Managing editor Robert Castillo took good care of the copyediting while respecting my voice. In particular, I am grateful to executive vice president and publisher Jamie Raab for creating a Web site for the book and allowing us to use far more photographs than originally planned. She has been by my side through several books and been a real support and friend.

Friends and Family: As I travel around the world I am supported—and often nurtured—by wonderful friends. I cannot thank them all—there are too many. But I have special words of gratitude for Michael Neugebauer and Tom Mangelsen. Tom has not only provided magnificent photographs but also introduced me to Ernie Kuyt and to the team that saved the black-footed ferret from extinction. I truly value the hours Tom and I have spent discussing endangered species and conservation, and enjoying the beauty of the wild places. You can see his amazing photography at www.mangelsen.com.

I could never have gotten through the months and years of writing this book without Mary Lewis, staunch companion of the road. Mary was the master coordinator of my crazy schedule, working miracles to ensure I got to fly with the cranes, spend the night with the ferrets, and meet the countless heroes described in *Hope for Animals and Their World.* And then, of course, there is her sense of humor. What a friend she is. It is sad she is not here as I type the last words of this marathon, but recovering from hip replacement surgery in the UK.

My crazy schedule and the need to spend all free moments wedded to the book has meant that I had less time than usual for my son and grandchildren. I thank them for their understanding. And a very special thank-you to my very special sister, Judy. If she wasn't there in the Birches I would have had no bolt hole in which to hide and to write, between trips around the globe. Judy, with her quiet common sense and strong support, was my anchor in the storm.

Foreword

JANE'S FEATHER
By Thane Maynard

The idea for a book of hopeful stories about wildlife was launched on an autumn evening in 2002. In the middle of a public lecture at a sold-out basketball arena, Jane stepped away from the podium and said her classic line, "Let me tell you a story . . ."

Reaching behind the podium, Jane slowly pulled out the largest feather I'd ever seen; indeed, one of the largest feathers in the world. It was a primary feather from a California condor, the most endangered animal in America. She told the enthralled gathering that she carried it with her for inspiration because it reminded her not that magnificent creatures were disappearing—as is so often reported, even to children—but instead that many species are coming back from the brink of extinction. Thanks to the hard work of a great spectrum of experts, activists, students, and enthusiasts, the California condor is flying again.

When her lecture was through, Jane walked up the stairs through the cheering crowd with the feather held aloft like the symbol of a tribal chief. Indeed, in that moment on such a fair fall night, we six thousand gathered there were a tribe united to care for wildlife and the natural world around us. After all, we had learned, such diversity is what holds the earth steady.

This book is a starting point to share the hope of such a dream. A dream in which caring people of all ages, from all over the world and all walks of life, show that it is possible to help, rather than harm, the rest of the world around us. For it is not counter to human nature to be hopeful. In fact, it is quite the opposite—it is essential to our nature.

People are as persevering as gray squirrels after a bird feeder and as tenacious as the termites who rebuild the topsoil on the forest floor. And just as nature has evolved to be nearly immeasurably resilient, filling in gaps created by storms, disease, and other calamities, so have human beings, both as individuals and as cultures, proven the ability to come back from disaster time and again. This is perhaps our greatest strength. As British author John Gardner put it, "We are at our best when the way is steep."

I really have no idea why Jane and I are so disproportionately buoyant in such a time of loss. I've even been called "a public nuisance" because my NPR radio broadcasts, *Field Notes with Thane Maynard* and *The 90-Second Naturalist*, promote a sense of wonder about nature rather than a sense of gloom. And while I know that we live lives of unprecedented destruction, I am blessed to also know many great people effectively working (and most of them quietly) to save what they can. To me they are like Nelson Mandela and Martin Luther King, carrying on with their miracles that many others believe impossible.

It is this same sort of passion that is represented in nearly every effective conservationist I have ever known. While the naysayers stand by wheezing and huffing and puffing about how "this will never work," or "it's too late to save this species or habitat," or "be practical, we have to compromise with the developers," it is the truly passionate conservationists who *never* give up. They are empowered by the hard work. You can see it in their eyes.

Maybe I'm also optimistic because in many countries, I find a growing sense of pride in their flagship species and their natural heritage. Just as importantly, there is a sense that they have a reason to protect what is still there. Not only because it is good for tourism or foreign exchange, but also because it is important to them and their children.

So today, when we live in an age of terrible loss all around us, rather than sadness for what we've done, it is essential that we express hope for what can be done. In order to do that, we need guiding lights—role models—who can light the way. For there are thousands of success stories of wild animals and plants that are making a comeback. And of people who are helping to protect the natural world we depend upon. They are, as Martin Luther King described himself in his self-scribed eulogy, "drum majors" for wildlife conservation.

And speaking of role models, it's worth noting that while we were pulling together this collection of conservation success stories, almost every conservationist we talked with harkened back to the key role

Jane's early work had in shaping their careers. Some mentioned the 1960s cover stories in *National Geographic* magazine. Others referenced the early TV specials about her life among the wild chimpanzees. And almost everyone spoke of the direct impact of Jane's seminal research chronicled in her 1971 book, *In the Shadow of Man*. The significance of her first book to these modern-day conservationists encompassed much more than just Jane's scientific accomplishments.

As Dr. David Hamburg, of the Stanford University School of Medicine, wrote in the original forward to *In the Shadow of Man*, "Once in a generation, there occurs a piece of research that changes man's view of himself. The reader of this book has the privilege of sharing such an experience."

At the time, of course, he was marveling at Jane's remarkable discoveries in chimpanzee behavior. However, her long-term study of wildlife, the first of its kind, also changed the way men and women viewed the possibilities of their own lives and careers. For there is not a "field biologist," as the new parlance goes, who does not owe a debt to the inspiration of Jane Goodall.

And now, nearly half a century in, Jane's ongoing work has motivated *two* generations of researchers and conservationists, including the people in this book working tirelessly to save wildlife. This group covers a wide spectrum. Some were educated at the best universities in the world. Others are largely self-taught through a lifetime of working with animals. Most are broke, since nobody goes into protecting wildlife for the money or vacation time. Group members span from their twenties to their seventies in age; some of them are politically savvy, and others are obstinate. But they share two things in common: They refuse to give up or take no for an answer, and they recognize that Jane Goodall authentically understands the relationship that is essential between wildlife and humans.

These are their stories.

Introduction

I am writing this from my home in Bournemouth, England. I grew up in this house, and as I look out my window I can see the very same trees I climbed as a child. Up high in those trees I believed I was closer to the birds and the sky, more a part of nature. Even as a very young child, I felt most alive in the natural world, and almost every book I read—borrowed from the local library—was about animals and adventures in wild untamed places in the world. I began with the stories about Doctor Dolittle, that English doctor who was taught animal languages by his parrot. Then I discovered the books about Tarzan of the Apes. Those two books inspired a seemingly impossible dream—I would go to Africa one day and live with animals and write books about them.

Perhaps the volume that influenced me most was called *The Miracle of Life*. I spent hours poring over the small print of those magical pages. It was not a book written for children, but I was absolutely absorbed as I learned about the diversity of life on earth, the age of the dinosaurs, evolution and Charles Darwin, the early explorers and naturalists—and the amazing variety and adaptations of the animals around the world. And so, as I grew older and learned more and more, my love of animals broadened from my hamster, slow worm, guinea pigs, cats, and dogs, to a fascination for all the amazing animals I read about in those books. There was no television when I was young: I learned everything from books—and nature.

My childhood dream was realized when I was invited to Kenya by a school friend. I set off when I was twenty-six years old, after working as a waitress to save the fare. I went by boat because it was cheapest, calling in at places I had read about such as Cape Town and Durban, and finally arriving in Mombasa. For me it was especially exciting to arrive at the Canary Islands—for Doctor Dolittle had been there, too! What adventure, back then, for a young woman traveling alone.

Once I reached Kenya, my love of animals led me to Louis Leakey, who eventually entrusted me with the task of uncovering the secrets of the behavior of the animal most like us. (Quite extraordinary when you consider I had no degree and back then girls did not do that sort of thing!) That study of chimpanzees, in Tanzania's Gombe National Park, has lasted for half a century and helped us understand, among other things, more about our own evolutionary history. It has taught us that the similarities in biology and behavior between chimpanzees and humans are far greater than anyone had supposed. We are not, after all, the only beings with personalities, rational thought, and emotions. There is no sharp line dividing us from the chimpanzees and the other apes, and the differences that obviously exist are of degree, not of kind. This understanding gives us new respect not only for chimpanzees, but also for all the other amazing animals with whom we share this planet. For we humans are a part of, and not separate from, the animal kingdom.

We are still studying the chimpanzees of Gombe, and I might well have stayed there, with the animals and forests I love, if I had not attended a conference called Understanding Chimpanzees. It was that conference, in 1986, that changed the course of my life. Field researchers from all the study sites across Africa came together for the first time. There was one session on conservation that was utterly shocking. Right across their range, the chimpanzees' forests were being felled at a horrifying rate, they were being caught in poachers' snares, and the so-called bushmeat trade—the *commercial* hunting of wild animals for food—had begun. Chimpanzee numbers had plummeted since I began my study in 1960, from somewhere over a million to an estimated four to five hundred thousand (it is much less now).

It was a wake-up call for me. I went to the conference as a scientist, planning to continue working in the field, analyzing and publishing my data. I left as an advocate for the chimpanzees and their vanishing forest home. I knew that to try to help the chimpanzees, I must leave the field and do my best to try to raise awareness and hope that we could start to halt at least some of the destruction. And so, after spending twenty-six years of my life doing what I loved best in the place I loved best, I took to the road. And the more I traveled around the world, giving lectures, attending conferences, meeting with conservationists and legislators, the more I realized the extent of the devastation we are wreaking on our planet. It was not just the forests harboring chimpanzees and other African animals that were endangered—it was forests and animals everywhere. And not only forests, but all of the natural world.

Life on the road is hard. Since 1986, I have traveled some three hundred days a year. From America and Europe to Africa and Asia. From airport to hotel to lecture venue; from schoolroom to corporate conference room to government offices. But there are some perks along the way. I get to visit some incredible places. And I get to meet some truly wonderful and inspirational people. And I hear, among all the terrible news of the ongoing destruction of the natural world, some stories of people who have prevented the felling of an old-growth forest, stopped the building of a dam, succeeded in restoring a despoiled wetlands, saved a species from extinction.

Even so, evidence is mounting of a sixth extinction—this time caused by human actions. To keep up my spirits when I was tired and things seemed extra-bleak, I made a collection of what I call my "symbols of hope." Many illustrate the resilience of nature—such as a leaf from a tree found in Australia, previously known only from fossil imprints on rocks. A tree that has survived seventeen ice ages and is still alive and well in a hidden canyon in the Blue Mountains. A feather from a peregrine falcon that was flying again in an area where it had been locally extinct for a hundred years and another from a California condor, a species rescued from the brink of extinction. This was what caught Thane's attention when I was lecturing at the zoo in Cincinnati. He said I should write up those stories. I told him I intended to—but there was so little time. He said he would help. Thane is a kindred spirit. He, too, is filled with optimism for our future.

Clearly this is a very different book from the slender volume originally planned. I kept meeting amazing people who had done amazing work to prevent animals from becoming extinct. And I met them all over the world. How could I write about the California condor and not the whooping crane? And what about the giant panda, symbol of conservation? Then, somehow, word got out that we were writing this book and information flooded in—why were we not including insects? Amphibians? Reptiles? And surely the plant kingdom was important, too?

And so the book grew, not only in volume, but also in concept. It seemed so important to discuss some of the species believed extinct that have been rediscovered—sometimes more than a hundred years after they had been written off. And to write about the wonderful work being done to restore and protect habitats. I found that people got really excited about the idea of sharing the good news, shining a light on all the projects, large and small, that together are gradually healing some of the harm we have inflicted. It has been several years in the making, this book, and it has taken me on a fantastic journey of exploration: I have learned ever more about animal and plant species brought to the brink of extinction by human activities and then—sometimes at the very last minute and against all odds—been given a reprieve. The stories shared here illustrate the resilience of nature, and the persistence and determination of the men and women who fight—sometimes for decades—to save the last survivors of a species, refusing to give up.

There is Old Blue, at one time the very last female black robin in the world who, with the help of an inspired biologist, saved her species from extinction. There is the individual tree, the very last of its kind, that, having been almost eaten to death by browsing goats, was killed by a forest fire—yet found the energy to produce seeds on its last living branch. With the help of inspired horticulturists, the species sprang back, like the phoenix, from the ashes.

It is these and many other human and other-than-human heroes that you will meet in the following chapters. There are tales of adventure and high courage, as biologists risk their lives to climb sheer rock faces or leap from wildly tossing boats onto jagged rocks, and pilots maneuver helicopters through forbidding landscapes in terrible weather. There are stories of men and women brought close to despair as they battled bureaucracies to try to save a species from extinction, knowing that delay caused by human obstinacy was lessening their chances of success with each passing day. There is an account of a man trying to persuade a falcon to copulate with his hat and another who mimics the courtship dance of a crane to persuade her to lay an egg.

Many of the rescue programs are ongoing even as we write. New generations of whooping cranes and northern black ibises are still being taught new migration routes, led by human devotees in flying machines. New breeding and release techniques for giant pandas, and better protection of wild habitat, offer hope for their future in China, but there is a long way to go. The plight of the Asian vultures that died in their hundreds of thousands from non-intentional poisoning is being addressed through captive breeding and "Vulture Restaurants" in the wild, but there is much, much work to do.

We realize that there are countless other programs going on around the world to conserve existing populations of animals and plants. But we had to pick and choose, and we included mainly stories that we knew about, firsthand. I wish we could include the efforts of the pioneer conservationists, such as Theodore Roosevelt, who established the first national parks and reserves for the protection of wilderness areas.

Or write about the farsighted people who worked to protect the last of the beavers from an industry desperate to plunder their pelts for the making of hats. There are many who have fought to save other mammal and bird species from extinction because of our insatiable desire to bedeck ourselves with their skins, furs, and feathers. Koala bears might no longer be with us but for those who realized, back in the 1800s, that they would soon be gone if steps were not taken to save their eucalyptus forests. Indeed, there are countless species not even classified as endangered today that might well have become extinct were it not for caring people who protected them long ago. To those early pioneers in conservation we owe a great deal.

In October 2008 in Barcelona, Spain, the International Union for the Conservation of Nature (IUCN) released results of a global survey of mammal populations. It concluded that "at least a quarter of mammal species are headed toward extinction in the near future." And tragically, for many, there may be little that can be done. Yet I have been so inspired by the stories included in this book and by the people who refuse to give up.

There is an old maxim: "While there is life, there is hope." For the sake of our children we must not give up, we must continue to fight to save what is left and restore that which is despoiled. We must support those valiant men and women who are out there doing just that. And it is important for us to realize that we cannot relax our efforts on behalf of endangered animals—for the threats to their survival are ever present, often growing. Human population growth, unsustainable lifestyles, desperate poverty, shrinking water supplies, corporate greed, global climate change—all these and more will, unless we are vigilant, undo all that has been accomplished.

It is inevitable that more and more species will need a helping hand if they are to continue to share the planet with us. So it is fortunate that increasing numbers of people are waking up, becoming aware of the damage we are inflicting on the web of life, and wanting to do their bit to help, whether as wildlife biologists, government officials, or concerned citizens.

One thing is certain—my own journey of exploration will not stop. I shall go on collecting stories, meeting and talking with more extraordinary and inspirational people. There are many to whom I have only spoken on the telephone, but now I want to meet them: I want to look into their eyes to see the spirit of determination that keeps them going, and look into their hearts to glimpse the love for the species or the natural world that takes them to lonely, all-but-inaccessible places. And I want to share their stories with young people around the world. I want them to know that, even when our mindless activities have almost entirely destroyed some ecosystem or driven a species to the brink of extinction, we must not give up. Thanks to the resilience of nature, and the indomitable human spirit, there is still hope. Hope for animals and their world. And it is our world, too.

—Jane Goodall, February 2009

PART 1

Lost in the Wild

Introduction

Children are fascinated by dinosaurs. I used to imagine myself transported into the past, my imagination stimulated by Jules Verne's *Journey to the Center of the Earth*. In my mind I would roam those ancient landscapes with the giant vegetarian brontosaurus, unharmed by the mighty tyrannosaur. I loved, too, mind-walking in the older world of the giant amphibians, that watery realm of swamps and huge ferns. And sometimes I dreamed of watching woolly mammoths and saber-toothed tigers. But they were gone, and I had no time capsule. And there were no marvels of technology to re-create those creatures of long ago—as did the extraordinary BBC TV series *Walking with Dinosaurs*.

And then I learned, from one of my books, about the dodo. That extinction was very different from the loss of the dinosaurs. The dodo (and countless others) would still have been around, I discovered, but for modern *Homo sapiens*. Of course, our Stone Age ancestors had hunted and killed animals. I would later see evidence of this when I worked with Louis Leakey in Olduvai Gorge. But it was hard work for them with only their primitive stone tools. Moreover, the prey animals in Africa had evolved along with the predators that hunted them, and had developed myriad ways to escape being killed. How different when Captain Cook and his sailors killed the unsuspecting flightless dodos, feeling safe on their island with no instinct for flight—and so they were eaten to extinction.

When I was a child, more than seventy years ago, there was no television and no Internet to trap me in front of electronic screens. Instead I spent hours watching birds and insects in our garden, and reading books.

Back then most of the animals that are so endangered today lived safely in as-yet-unlogged forests, undrained wetlands, and unpolluted fields and oceans. Yet even then, of course, large-scale slaughter of wildlife was taking place. The American bison herds were being decimated, wolves were being exterminated, and animals in their hundreds of thousands were being trapped and killed for their skins, their fur, their feathers—and for specimens to stuff for natural history museums. Big-game hunters were "bagging" and bragging about trophies. And passenger pigeons were hunted to extinction. For the most part, no one thought much about any of that, and anyway nature's natural resources, to most people, seemed inexhaustible.

But gradually our human populations have grown, and the destruction of the natural world has intensified. One after the other, more and more of the extraordinarily varied life-forms of our planet have joined the dodo and the passenger pigeon. Mostly they are small creatures and plants, often endemic to a particular area of rain forest or other habitat that has been destroyed. But fish and birds have gone as well. And Miss Waldron's red colobus was pronounced extinct in Ghana at the end of the last century. So much has gone even during the seventy-five years since I was born.

Will a nature-loving child born seventy-five years from now long to see a live elephant as I longed to see a woolly mammoth? Will she wish desperately for a time machine in order to experience a real rain forest and watch orangutans and tigers? Will she yearn to know a lost and mysterious deep-sea world of the great whales? And if, in seventy-five years, these animals exist only in digital libraries or as dusty museum specimens, how will she feel?

When I was a young girl, it was possible for me to forgive Captain Cook and the people of his era, for they had no idea of the direction we were heading (though they were unknowingly mapping out the path of the future). But at that time, the world was largely unexplored, its wonders undiscovered—and there were far fewer human beings. Still, if a child seventy-five years from now finds that most animals have gone from the earth, she will not be able to excuse the behavior of those who destroyed them. For she will know that they were lost not from a position of ignorance, but because the majority of humans simply did not care.

Fortunately, some people do care a great deal, and sometimes heroic efforts are being made to save and conserve threatened and endangered species. But for them, the list of extinct animals today would be much longer. I have been privileged to meet many of them, and in this book I look forward to introducing as many as I can, along with the animals, plants, and habitats to which they have devoted their lives.

The stories we are sharing in the first two parts show how complicated a business this conservation of wildlife is. For it is necessary to integrate research, protection in the wild, habitat restoration, captive breeding, and raising awareness in the local population. And there are restrictions—everything must be undertaken under the watchful eyes of government authorities. Also, it is inevitable that when passionate people with different perspectives try to work together, differences of opinion arise, and these opinions will be hotly defended—and although, through discussion and compromise, agreement will usually be reached, a good deal of time and effort may be wasted along the way. In the best-case scenario, organizations working to protect an animal and its environment cooperate for the good of the species, and the public volunteers its help.

Part 1 tells the stories of six mammal and bird species that actually became extinct in the wild. They were saved only through captive breeding with the goal of returning their progeny to the wild once their numbers had increased and areas of habitat had been set aside for their lasting protection. But the issue of captive breeding was—and still is—highly controversial. There are objections to such projects from those who feel last-minute solutions will not work, and are a waste of time and above all money. Fortunately the passionate biologists who worked to save the six species in this section refused to listen to them.

I have fallen in love with black-footed ferrets. Tiny in size, mighty in courage, and utterly enchanting, they have been brought back from the brink of extinction by a team of dedicated and inspired biologists. For in the brilliant emerald of the ferret's nighttime eyes lies hope for the future of the great North American prairies. *(Jessie Cohen, Smithsonian National Zoo)*

Black-Footed Ferret
(Mustela nigripes)

In the Lakota culture, the black-footed ferret is called *itopta sapa*: *ite*—face, *opta*—across, *sapa*—black. The Lakota admired *itopta sapa* for its cunning and elusiveness and held it sacred. Creatures that were hard to kill, like *itopta sapa*, were thought to be protected by the earth power and the thunder beings. Today the Lakota still consider this ferret sacred.

At one time, short- and mixed-grass prairies, home to the black-footed ferret, covered nearly one-third of North America, from Canada to Mexico. This vast area was also home to the great bison herds as well as the prairie dogs that lived in huge colonies, and provided food and homes for the ferrets, who lived in their burrows.

When Europeans arrived in North America, things began to change. Human developments transformed the prairies, so that more and more prairie dog habitat was destroyed, and the ranchers began their ongoing campaign to poison as many as possible. They maintained that the rodents competed with their livestock for grass and that their burrows would cause broken legs. By 1960, using the most conservative calculations, prairie dogs had lost some 98 percent of the land they had once occupied. New diseases were also brought to the prairies: Sylvatic plague, for example, entered North America around the turn of the century and is having a devastating impact on prairie dog towns to this day.

Prairie dogs, being rodents, can quickly bounce back from a population decline, but not so black-footed ferrets. They are predators with a naturally low population that is spread out over a wide area. As their numbers declined, it became more and more difficult for them to replenish themselves.

Disappearing into Extinction

In 1964, the federal government was actually debating whether these wild ferrets should be listed as extinct when a small population (only 20 of the 151 prairie dog colonies in the area were occupied) was discovered in Mellette County, South Dakota. As time went on, however, it became clear that this small population was decreasing, probably because of fragmented habitat and the poisoning of prairie dog colonies.

In 1971, six of the Mellette County ferrets were captured to form the nucleus for a captive breeding program. Tragically, four of these precious lives were lost when they were vaccinated against distemper, even though the vaccine had not harmed the Siberian ferrets on which it had been tested. Three more were then captured, but the program seemed doomed. Over the next four breeding seasons, one of the captive females refused to mate, and although the other twice produced litters of five, each time four of the five were stillborn, and the fifth died soon after birth. Meanwhile, the wild ferrets of Mellette County were disappearing—the last time one was seen was 1974.

I can imagine the desperation of the team working on the captive breeding as they watched the species falling into extinction. In 1979, the last remaining captive black-footed ferret died of cancer, and the federal government again debated listing the species extinct.

A Fateful Encounter

And then, on September 26, 1981, two years after the death of the last captive black-footed ferret in South Dakota, something very exciting happened. In Meeteetse, Wyoming, on the property of John and Lucille Hogg, a small animal got too close to Shep, their blue heeler ranch dog, when he was eating his dinner—and Shep naturally killed it. John found the strange-looking animal by Shep's dish and tossed it over the yard fence, but when he told his wife about it, she became curious and retrieved the body. She was enchanted by the beautiful little creature, and took it to the taxidermist to be preserved. And the taxidermist recognized a black-footed ferret!

A group of excited ferret enthusiasts quickly gathered to survey the area. How excited Dennie Hammer and Steve Martin must have been when they saw two emerald-green eyes shining as a little head popped up from a burrow—vindication at last for their conviction that wild ferrets still existed! Yet only pure luck had provided this proof. Over the next five years, private, state, and federal conservation biologists and many volunteers worked to learn more about the ferret population. They searched for the ferrets with spotlights, trapped them and marked them with tags, fitted them with tiny radio transmitters on neck collars (so the team could spy on the ferrets' nocturnal habits), and used a new technology, tiny transponders that could be implanted in the neck (which allow short-range identification of an individual animal).

"None of us took them for granted," Steve Forrest, a team member, told me later. "We knew the ferrets as individuals. We lived with them. We knew these were the last members of the species."

My Night with the Ferrets

In April 2006, thanks to my friend Tom Mangelsen, the photographer, I met some of that original dedicated team—Steve and Louise Forrest, Brent Houston, Travis Livieri, Mike Lockhart, and Jonathan Proctor. We gathered in Wall, South Dakota, at Ann's Motel. I soon found that this would be an all-night experience, for the ferrets are not active till around midnight. We set out in the evening, stopping for a picnic to watch the sun set behind the extraordinary rock formations of the Badlands, bringing out the fantastic colors—ocher, mauve, yellow, gray, and all the subtle shades between.

Gradually, as we drove toward the prairies, the day faded until all color was drained from the landscape. There was no light pollution apart from the headlights of our trucks, and the stars were large and brilliant in the wide sky. It was strange to think that we were driving over the thriving underground prairie dog towns—that were home, too, to the black-footed ferrets.

It was close to midnight when Brent called out: "There's one!" And I saw the eyes of a small animal shining brilliant emerald green as they reflected his spotlight. As we drove closer, I made out the ferret's head as she looked at us, listening to the engines. She did not vanish as we cautiously drove closer. And when she did duck down, she could not resist popping up for another look before disappearing. When we eventually went over to peek down the burrow, there was her little face, peeking back at us, not at all afraid. Travis later returned to take a reading of her transponder chip—which is how I know she was female.

Travis, who was in a second truck, found another ferret—a male—who soon darted into a burrow. It was the time of year, Travis explained, when males check out the burrows looking for females in estrus (in heat). Sure enough, after a while the ferret bounded out and raced to another burrow. He moved like lightning, his tiny body stretched out long and thin. We followed. Obviously, no suitable female there, for soon he reappeared, stood upright to look around, and stretched tall as he could—checking for coyotes and foxes. Then he streaked off and vanished into yet another burrow. That burrow was apparently female-less also, for he soon emerged again. During his next cross-country run, our ferret bumped into—physically bumped into—a horned lark! As the startled bird flew up, the ferret did a complete backflip to land, like the acrobat he is, on all four feet facing the way he was going before. Without a pause he raced on toward the next burrow. It was a fabulous show! I doubt anyone has ever seen a black-footed ferret–horned lark encounter of that sort.

How Bureaucratic Obstinacy Nearly Led to the Extinction of the Ferrets

The next day, Tom and I were able to sit down with Travis, Steve, and Jonathan (the others had to leave) and talk about the black-footed ferret recovery program. Steve described the harrowing events that took place four years after the miraculous discovery of the wild Meeteetse ferrets. In August 1985, they got permission to assess the status of the ferret population, as they had done each year. They found 58 individuals, a marked decline from the 129 found the previous summer. In September, they estimated there were only thirty-one, and by October the wild ferrets were down to just sixteen.

The biologists believed that the ferrets had been afflicted by distemper, and they sought permission from the Wyoming Game & Fish Department (responsible for the black-footed ferret program) to capture some individuals so they could get blood samples for veterinary testing. Permission was refused on the grounds that the procedures were too invasive. The situation worsened—it became clear that the juveniles were not surviving.

Brian Miller, whom I met later, was part of the team at that time. "Walking the area was not like previous years, when ferrets reliably occupied areas," he told me. "Now you would see a ferret in his or her territory on one night, and the next night that area was empty." This situation, while desperately alarming to the biologists, was ignored by Wyoming G&F. Finally, a meeting was arranged to discuss the ferrets' plight. Steve, Louise, and Brent, along with other biologists, were all present, as were various staff of Wyoming G&F, a representative of IUCN, and a group of old-time game rangers who had no understanding of—or patience with—conservation biology.

At this meeting, the scientists were criticized for not providing good data—data about the suspected distemper epidemic that they had not been allowed to collect! The discussion became heated. The scientists stressed the urgency of trapping more ferrets for intensive captive breeding. Permission was again refused. Things were going badly for the researchers and, thus, for the future of the ferrets when the Wyoming G&F veterinarian came into the room, clearly agitated.

At that time, there were six ferrets in captivity, trapped earlier for the captive breeding program that had, after prolonged pressure from many sources, eventually been agreed to by Wyoming G&F. One of the six, reported the veterinarian, had died, and another was very sick. The reason—distemper, almost certainly contracted in the wild. "All at once it was very quiet," said Steve, flashing a broad smile as he recalled the discomfort of their obstinate adversaries. At last, the scientists had their evidence.

Gone in the Wild

Yet even then, they were only allowed to catch animals from the central part of the range—leaving the most vulnerable individuals in the peripheral areas to disappear, lost forever. And despite the fact that the ferrets were clearly on the brink of extinction, Wyoming officials did not deviate from a planned strategy—only six more ferrets (the original six were dead or dying) could be caught. And they could only trap one per day—because that was the rate at which cages were being constructed. Offers to bring in a company to make them faster were ignored.

"We started right away," Steve told me. Over the next three nights, they covered forty square miles of prairie, trapping ferrets in a desperate attempt to save the species. On the third night, Brent had just trapped two when an officious local game officer arrived and told him he had exceeded his quota. "He told Brent to release one of the two," said Steve, "and Brent refused." They practically came to blows as the game officer simply cut the trap open.

By that time there were so few ferrets, and Wyoming G&F had been so uncooperative, that there had been little choice as to which individuals were trapped. Thus the nucleus of the breeding group was three adult females and one juvenile (Emma, Molly, Annie, and Willa), as well as two juvenile males (Dexter and Cody). A specialist in captive breeding warned that without an adult male the onset of breeding would be delayed, but Wyoming G&F ignored this advice, and though an adult male was seen in a peripheral area, his capture was not permitted. Thus there were no litters in the captive group the next season.

It was an agonizing time. Brian Miller, who had paired the captive ferrets, told me how they had watched the breeding cages on a remote camera all night. "We were wondering," he said, "if we were watching the modern version of Martha, the passenger pigeon." Martha was the last individual of a species that is now extinct. She died of old age in a zoo and is now mounted in the Smithsonian. "I once went to see her," said Brian. "Was that to be the fate of Emma, Molly, Annie, Willa, Dexter, and Cody, too?"

By the following summer, 1986, it seemed that only four adults—two males (Dean and Scarface) and two females (Mom and Jenny), each of whom gave birth—were left in the wild. Now, *finally*, Wyoming G&F agreed that all four adults and the eight remaining juveniles should be trapped for the breeding program.

The biologists worked hard for the rest of the summer, and eventually the last ferret was captured—Scarface. At this point, eighteen captive black-footed ferrets, a handful of biologists, and an unproven captive breeding program were all that was left to buffer the species from extinction. Despite the fact that discord and bad feelings continued to plague the program, the ferrets began to breed, and gradually other centers were established across the country, so that the outbreak of disease or some other disaster at one facility would not wipe out the entire captive population.

Hard Versus Soft Release

Next, the arguments began over when and how ferrets should be reintroduced into the wild. The most acrimonious argument concerned the pros and cons of "hard release" (when animals are taken straight from the cage and let loose, usually with some food provided for a while) versus "soft release" (when the animals are given a variety of opportunities to gradually get used to a new life in the wild). Many of the field biologists felt strongly that it was not ethical to suddenly dump ferrets from small cages into the dangerous world of the prairies with no experience or training, but in 1991 the first forty-nine captives were hard-released into Wyoming's wilderness.

The next release site was Conata Basin in South Dakota, where I had met my first ferrets. Later I would meet Paul Marinari, who told me about one night he will never forget. He was searching for ferrets with Travis and four other biologists, spread out over the prairie.

Suddenly his radio sprang to life and a message "crackled through the South Dakota night proclaiming that multiple ferret eye-shine was detected from one burrow. This signified the first observation of a wild-born ferret litter (from captive-born parents) in the state. Those moments were goose bumps on goose bumps!"

Eventually, it was proved conclusively that hard release is not the best option—not only does soft release lead to much better short-term survival rates, but more individuals live to breed the following season as well. Gradually, more and more released ferrets survived. It had been established that they could be bred in captivity and that they could survive and breed in the wild. But could their habitat be preserved?

Paul Marinari releases a black-footed ferret into a preconditioning pen before its final journey into the wild. *(Ryan Hagerty)*

Saving the Prairies

During my visit with the team, as I came to understand the challenges they faced, I was interested to talk further to Jonathan Proctor about his work with the prairie dogs and the prairie ecosystem. Jonathan explained that one of the main problems for conservationists is that almost no rancher has a good word for prairie dogs. I met one of these old-timers as he drove by Ann's Motel. The prairie dogs, he said, were a real nuisance. There were all those holes in the ground that caused cattle and horses to break their legs. And, he said, the prairie dogs competed with the herds for the new grass. While no one I talked to had actually encountered any cows or horses with broken legs on the prairie, I listened to his point of view and respected what he had to say. I said it was a shame there wasn't some way around the problem without poisoning those cute little animals.

"Best prairie dog is a dead one," he said—but he reached out and touched my arm, as though he knew what I meant, and told me he'd watched my shows and thought I did a great job. It is so important to talk with people and listen to their point of view, to try to find solutions that will work for everyone. For this conflict between people and wildlife gets ever more intense as our human populations multiply and more and more wild land is taken over for development.

Perhaps, in the end, tourism will save the great American prairies, along with all the fascinating life-forms that make up the ecosystem. And the last of the old-time ranchers can offer visitors a taste of the old days, staying in an old-style homestead on land where, once again, bison roam. And where the Central Plains Indians (such as the Lakota and the Sioux), so much a part of the great prairies, and who are even now helping with restoration projects, will have a major role to play.

A Very Special Ferret

On the last morning of my visit to Wall, South Dakota, we gathered for breakfast, not wanting to part. How much I had learned, how complex the issues were, and how many challenges lay ahead. Before we said our good-byes, Travis told me about one of the individuals who had made a major contribution to the program. She was known, simply, as No. 9750 (the 97 indicates the year she was born). In 1996, Travis had released thirty-six captive-born ferrets into the wild, and No. 9750's mother had been one of the only four to survive. No. 9750 was born the following year in the first cohort of wild-born black-footed ferrets in the Conata Basin. "Their future was uncertain," Travis told me. "But No. 9750 survived and prospered and became a founder of the black-footed ferret population that now numbers approximately three hundred adults and kits annually in Conata Basin." No. 9750 lived for four years, which is quite old for a wild black-footed ferret. She had produced four litters and raised a total of ten to twelve youngsters.

In October 2001, Travis came upon No. 9750. She looked exhausted after raising her last litter, emaciated and with thinning hair and deep-sunken eyes. Kneeling to look down at her in the burrow, he knew she would not see another spring. Listening to Travis, I was miles away from the breakfast table, with its empty plates and cups. I was out on the prairie, bleak with approaching winter, with this tough dedicated man who was talking softly, saying good-bye to a very small, very tired black-footed ferret. "I want to say thank you, honey. I know we'll not see each other again." I could tell, by his voice, that he was all choked up, but I could not see for the tears in my eyes.

Everything You Wanted to Know About Ferret Breeding but Were Afraid to Ask

In April 2007, I squeezed a morning out of my tour schedule to visit the captive breeding program at the US Fish and Wildlife Service (USFWS) National Black-Footed Ferret Conservation Center in Wellington, Colorado, home to about 60 percent (roughly 160 individuals) of the captive population. (The rest are scattered in various zoos.) There I had a wonderful reunion with Travis, Brent, and Mike, who are all working there, and met for the first time Dean Biggins and Paul Marinari, both of whom I had heard so much about.

It is important, Paul explained, to determine exactly when the male and female are ready for breeding, whether the male's sperm count is healthy, whether a female has been successfully inseminated, and so on. One three-year-old female was having a small amount of saline solution squirted into her vagina. Not far away, Paul encouraged a male to leave the lower portion of his housing and climb up a piece of black tubing into a small wire cage. Once the ferret was there, Paul demonstrated how to gently squeeze the scrotum, which needed to be firm. If it was, he would be anesthetized and subjected to electro-ejaculation.

Next, we looked through the microscope at the fixed sample from another male and saw the little sperm there. He was ready, anyway! The results of all these necessary but undignified procedures were displayed on charts pinned up on the walls—showing which female had bred with which male, which couples were really incompatible, how many offspring had survived, and which, from the genetic point of view, could be allowed to breed. Clearly, the program has been successful—since it was initiated in 1987, it has resulted in the birth of more than six thousand black-footed ferret kits.

There was one great moment when Paul opened the upper cage of a female who had given birth two days before and I got to peep in—one of the first people to see the five tiny kits pink, naked, and blind, curled up there. Paul told me he never tired of watching them change "from a pile of squirmy little worm-like beings to chattering kits at sixty days of age." Some of them would be selected as reintroduction candidates. "Then," he said, "they undergo the most dramatic events for any captive animal: release into a preconditioning pen and, hopefully, reintroduction to the wild."

Ferret School

Travis was the one who first told me about the "ferret school" that starts when a captive ferret mother and her kits are placed in a large outdoor area where prairie dog burrows are occupied by prairie dogs. It will be the home and hunting ground of the kits for the next several months before they are sent for release into the wild, usually with their mother. This experience—living in a prairie dog burrow and hunting prairie dogs as prey—is a critical phase in preparing them for life on the prairies.

"It's where the kits get to experience wind, rain, dirt, and all the outdoor sounds of the North American prairie—and ultimately live prairie dogs," said Paul. "When the kits are placed into these pens, I often wonder what they must be thinking. They often stand in wonderment at such a large enclosure (compared with their indoor cage setting). Then they immediately play follow-the-leader as they almost stumble over their dam, who leads them around the pen, going in and out of each prairie dog burrow opening. Eventually, they settle down, becoming more and more secretive until the day arrives when it's time to free them from their captive setting for life in the wild."

The Future of the Black-Footed Ferret

The goal of the Black-Footed Ferret Recovery Plan is to reintroduce the ferrets into all eleven states where they once lived. Since the start of the program in 1991, Dean told me, more than three thousand ferrets have been released at reintroduction sites in eight of those states (Wyoming, Montana, South Dakota, Arizona, Utah, Colorado, and Kansas, and also into northern Mexico). Several of the sites, including the one I visited in Conata Basin, have successfully established wild black-footed ferret populations. Releases have occurred on federal, state, tribal, and private lands, and the black-footed ferret recovery program now encompasses many partner agencies, organizations, tribes, zoos, and universities. Wyoming Game & Fish, despite some of its past shortcomings, has been an integral and important part of the ferret program, overseeing a large population of ferrets in the state.

Dean, as mentioned earlier, was part of the team that captured the last wild ferrets in existence during 1986–1987 in Meeteetse. One of these was a female they named Mom. Before they captured her, she left a little paw print in the soil outside her burrow, and Dean had made a cast of it. As I was getting up to go, Dean gave me a replica of that cast. I looked down at the tiny print and thought of that bitter time when the dedicated team, to try to save a species, took the last wild individuals into captivity, and I was moved almost to tears. On the back Dean had written:

"Mom" August 30, 1986
Meeteetse, WY.
One of the last 18 black-footed ferrets.
To Jane from Dean Biggins, Travis Livieri, Brent Houston, Paul Marinari, Mike Lockhart 4.25.07.

It is one of my most prized possessions and travels with me around the world.

When I visited Australia in October 2008, I had the honor of releasing this mala into her natural habitat of Alice Springs Desert Park. This canvas bag served as a "pouch" for the mala when she was being transported. *(Peter Nunn)*

Mala or Rufous Hare-Wallaby
(Lagorchestes hirsutus)

I met my first mala in October 2008, and had the joy of releasing the captive-bred animal into a large fenced enclosure where she could get used to living in the bush. It was Polly Cevallos, CEO of the Jane Goodall Institute (JGI)–Australia, who first told me the heartwarming story of the rufous hare-wallaby, usually known by its Aboriginal name, the mala. She put me in touch with Gary Fry, director of the Desert Park in Alice Springs, where the mala are being restored. Two years after a first phone call, I arrived in the place I had wanted to visit ever since reading Nevil Shute's *A Town Like Alice,* in the heart of the Australian continent.

It had been a scorching-hot day, but it was cooling off by the time we reached Gary's house. I was traveling with Polly and the director of JGI's Roots & Shoots program in Australia, Annette Debenham. We dumped our bags, said a quick hello to Gary's wife and son, and met Dr. Kenneth Johnson, who had set up the mala captive breeding program in the 1980s. Then we all set off for the enclosure. Two of the Desert Park staff were already there with the mala, invisible in a cloth "pouch." I sat on the dry grass, and the mala was gently placed on my knee.

Presently a small face peered out. Very slowly she emerged, hopped out of the bag onto the ground, and stopped right there, a couple of feet away from me, looking around. She was beautiful, a small, delicate kangaroo, with shaggy soft grayish brown fur tinged with red. Eventually, investigating her surroundings, she moved slowly away, though she did not go far. I noticed how her tail, hairless like that of a rat, trailed the ground behind her (Ken told me later that this is how the Aboriginals identify mala tracks in the bush). Soon we left her to settle into her new temporary home. Like many other Australian mammals, malas are nocturnal: She would be able to explore during the night and feel comfortable sleeping the next day. And indeed, we got the report next morning: She had eaten the food left out and was sleeping in the shelter set up for her.

That evening, during a wonderful dinner cooked by Gary's wife, Libby, Ken and Gary told me the story of the mala. At one time, there may have been as many as ten million of these little animals across the arid and semi-arid landscape of Australia, but their populations, like those of so many other small endemic species, were devastated by the introduction of domestic cats and foxes—indeed, during the 1950s it was thought that the mala was extinct. But in 1964, a small colony was found 450 miles northwest of Alice Springs in the Tanami Desert. And twelve years later, a second small colony was found nearby. Throughout the 1970s and 1980s, these two populations were studied and monitored by scientists from the Parks and Wildlife Commission of the Northern Territory. Very extensive surveys were made throughout historical mala range—but no other traces were found.

Ken told me something of the heartache of the team working with the mala during those years. At first it seemed that the little animals were holding their own. But then in late 1987, the first disaster struck: Every one of the individuals of the second and smaller of the wild colonies was killed. From examination of the tracks in the sand, it seemed that just one single fox had been responsible. And then, in October 1991, a wildfire destroyed the entire area occupied by the remaining colony, and all the mala died. Thus the mala really did become extinct in the wild.

How fortunate that, ten years before, Ken and his team had captured seven individuals that had become the founders of a captive breeding program at the Arid Zone Research Institute in Alice Springs. And that group had thrived. Part of this success is due to the fact that the female can breed when she is just five months old and can produce up to three young a year. Like other kangaroo species, the mother carries her young—known as a joey—in her pouch for about fifteen weeks, and she can have more than one youngster at the same time.

Working with the Yapa People

In the early 1980s, there were enough mala in the captive population to make it feasible to start a reintroduction program. But first it was necessary to discuss this with the leaders of the Yapa people (*yapa* is their name for "Aboriginal"). Traditionally the mala had been an important totemic animal in their culture, with strong medicinal powers for old people. It had also been an important food source, and there were concerns that any mala returned to the wild would be killed for the pot.

And so, in 1980, a group of key Yapa men was invited to visit the proposed reintroduction area. Many of them, including the principal owner of the "Mala dreaming," took some persuasion to make the 120-mile trip to the site, since he believed mala to be "all finished up." But he did come in the end, this knowledgeable old man, and shared his vast understanding of the species with the group. It turned out that they were all as concerned for the future of the mala as Ken and his team, and the possibility of a food hunt was not even mentioned. The skills and knowledge of the Aboriginals would play a significant and enduring role in the project.

Ken and his team went ahead and built a fifty-by-fifty-yard enclosure out in the desert, and twelve mala from the successful breeding program were moved there, given some time to get acclimatized, and then set free. One year later some were still alive, and thirteen more were released. Unfortunately, through a combination of drought and predation by feral cats, all of them were killed or disappeared.

After this, with the help of the local Yapa Aboriginals, an electric fence was erected around 250 acres of suitable habitat about three hundred miles northwest of Alice Springs so that the mala could adapt while protected from predators. By 1992, there were about 150 mala in what became known as the Mala Paddock and another 50 in the Alice Springs colony.

However, all attempts to reintroduce mala from the paddock into the unfenced wild were unsuccessful. Over a two-year period, a total of seventy-nine were released; all disappeared or were killed (the evidence pointed overwhelmingly to cats and some foxes as the culprits). And so the reintroduction program was abandoned: The Tanami Desert was simply not safe for mala.

Ken and his team now faced a situation where mala could be bred, but not released. In 1993, a Mala Recovery Team was established to set new goals for the program. First, the team concentrated on finding suitable predator-free or predator-controlled sites within the mala's known range. The first place selected was a new endangered species enclosure at Dryandra Woodland in Western Australia—an area where, before it had been converted into the "wheat belt," mala had been common. Initially, captive-bred animals would live in a large enclosure; as their population increased, selected individuals would be radio-collared and released into suitable conservation reserves or national parks in the area.

Finally all arrangements had been completed, and in March 1999 twelve adult females, eight adult males, and eight small joeys were sent off from the Mala Paddock on a very long journey. Early in the morning, they were loaded into a station wagon for a bumpy three-hour ride along bush tracks to the nearest airstrip. Here a delegation of Aboriginals had gathered to see them off, a mark of their intense interest in the mala program. From there the precious cargo traveled on a chartered plane to Alice Springs, on a regular commercial flight to Perth, and finally by truck to their final destination. They arrived about four in the afternoon, and were released into their new home at seven o'clock. I can just imagine how anxiously the bags were opened—how would the little animals have survived that tough day? But all was well. The mala at once started feeding on fresh alfalfa, then hopped off to explore their new home.

The second translocation of mala from the Tanami Desert, a few months later, was to Trimouille, an island off the coast of Western Australia. First it had been necessary to rid the island of rats and cats—a task that had taken two years of hard work. Finally the island was ready to receive the mala, and the Aboriginal traditional owners had given their blessing to the project even though it involved sending some of their totemic animals far away from their "dreaming home." Twenty females and ten males were selected for the long journey. Once again, all arrived safely.

Six weeks after their release, a team returned to the island to find out how things were going. Each of the malas had been fitted with a radio collar that transmits for about fourteen months, after which it falls off. The team was able to locate twenty-nine out of the thirty transmitters—only one came from the collar of a mala that had died of unknown causes. So far the reintroduction had gone even better than expected. Today there are many signs suggesting that the mala population on the island is continuing to do well.

Reintroduction to the Sacred Lands

During my visit to Alice Springs, Gary told me that his part in the story started with the plan to reintroduce a number of locally extinct species into the Uluru-Kata Tjuta National Park. The 1,142-foot-high Uluru-Ayers Rock is the most sacred place for the Aboriginals. With Polly and Annette, I had flown over the area, and been amazed by the sheer size of this huge outcrop of red rock surrounded, for miles in every direction, by the flat expanse of the Simpson Desert.

In 1999, parks staff and other biologists met with key members of the local Anangu people to discuss which species should be reintroduced into the Uluru area. As with the Yapa aboriginals, the mala had played an important role in Anangu culture, and there was a real wish that it be brought back.

"This little wallaby," Gary told me, "was the most preferred of all species for Anangu women and second most preferred for senior Anangu men." Gary also learned that even after the mala had disappeared from Uluru, the Anangu had kept their memory alive and strong, for the mala are an important part of the creation stories. Indeed, Gary told me that the loss of the little wallabies from Uluru had been of great significance for senior and powerful Anangu people, and brought them deep sadness.

Jim Clayton, an inspired park ranger based in the Uluru-Kata Tjuta National Park, worked with the Anangu to map out where the enclosure would be, encouraging them to help in the building and maintenance of the all-important fences that would protect the mala from introduced predators. And Gary tried to persuade the Anangu to set aside a large area of their tribal land. In a big-enough enclosure, he felt, the mala would be able to get on with it, needing little assistance from humans apart from maintenance of the fence.

Some time passed during which Gary heard nothing. And then, finally, Jim called: "We had some difficulty mapping the area," he said, "because we had to miss some sand dunes, and some stands of desert oaks . . . How does 170 hectares [approximately 420 acres] sound?"

It sounded great! An enclosure of this size was just the "fillip" (an Australian/British phrase for "boost") that was needed for the program, Gary told me. He felt very strongly about the outcome of the reintroduction not only for conservation of the species, but also for the conservation of the culture of the Anangu.

Six years later to the month, at 7 AM on September 29, 2005, twenty-four mala were released into the newly constructed, predator-free paddock in the Uluru-Kata Tjuta National Park. Many Ananga were present, and the press was well represented. It was a fantastic occasion, the culmination of years of planning and hard work.

Just as I was completing the manuscript for this book, I received an e-mail from Peter Nunn, a staff member at Alice Springs Desert Park. "I thought you would love to know that the Mala you released into our Free Range area at the Alice Springs Desert Park is doing really well," he wrote. "So well in fact that she has a young joey growing in her pouch! I was lucky enough to have her wander straight past me when I was spotlighting up there the other night, and she is looking wonderful. I hope that great news puts a big smile on your face!"

And, indeed, it did.

SURROGATE MOTHERING OF JOEYS
The Story of the Black-Flanked Rock-Wallaby
(Petrogale lateralis)

Just after I met my first mala, I also met my first black-flanked rock-wallaby at the captive breeding programs of Monarto Zoo, near Adelaide. Peter Clark, the senior curator, told me that as a result of environmental degradation and predation by and competition with introduced species, numbers of "warru" (to give the species its Anangu name) had plummeted to a low of only fifty to seventy individuals.

Then in 2007, a clever plan was implemented to try to save it—one that has been used very successfully to boost numbers of other endangered wallaby species. It is based on an unusual reproductive strategy: If a female wallaby loses a joey, she is able to replace it by activating a fertilized egg that she has stored internally. And so a team of biologists working in the field capture female warru, check their pouches, and if they find tiny, partially developed joeys, they "steal" them and take them by plane to Monarto Zoo, where they are implanted into the pouches of non-endangered yellow-footed rock-wallabies (*Petrogale xanthopus*). Because the stored "contingency" embryos soon start to develop in the wild mothers, there is no loss to the wild population.

Members of the local Anangu community accompanied the first twenty stolen joeys (each of which they had named) on the flight to Monarto. All survived the capture and journey and thrived in the pouches of their new mothers, Peter told me. But then, before they became too independent, they were taken away so that their rearing could be taken over by zoo staff. This was important, Peter said, since they will be used for captive breeding during which it will be necessary to check their pouches frequently, and this will be much less stressful if they are familiar with their human handlers.

Currently both the government and the local Anangu people are carrying out continuous monitoring of the warru population in three remnant rock-wallaby sites, using tracks, scats, and radio tracking of previously trapped individuals. While numbers are still low, it is encouraging that several new individuals have been found. These are the areas where the captive-bred warru from Monato will eventually be reintroduced once there are enough of them, and once predator control programs are working satisfactorily.

Before I left, one of the keepers, Mick Post, took me to meet a breeding female. She had arrived with the second group of joeys who had been named by zoo staff, and instead of having an Aboriginal name she was called Maureen! She was enchanting—an elegant-looking animal, about one and a half feet tall when she sat upright. Her fur was dark gray with blackish stripes on her face and flanks. She was completely at ease with us—and when I sat on the floor of her enclosure she climbed onto my knee and just sat there, looking around with interest at the cameras pointed at us. Mick said that she sometimes sits on his head as he is cleaning her enclosure, watching all that is going on. It was a real privilege to meet the dedicated team working to ensure the survival of Maureen and her relatives and descendants.

Captive-bred California condors released into the wilds of Baja California, Mexico. (*Mike Wallace*)

California Condor
(Gymnogyps californianus)

The California condor is one of the biggest birds in North America, weighing up to twenty-six pounds and standing nearly a yard tall, with a wingspan of nine and a half feet. As a child, I knew only about the vultures of Africa and Asia, for they frequently figured in my storybooks—usually in a somewhat sinister role as they patiently watched the hero, close to giving up as he struggled across the desert, thirsty and wounded. But one look at their hooked beaks, sharp talons, and cold greedy eyes, and he would summon the strength to reach safety. During my years in Africa, I have spent many hours watching the fascinating behavior of those vultures in the wild, but the California condor, which I learned about much later, I have seen only in captivity.

Initially, I was not attracted by its appearance. The bare skin of the head is so—well—bare! And its redness is the color of a boiled lobster. Truly, the condor is one of nature's odd experiments, where so much of the poetry, so much of the magic, went into the fashioning of those glorious wings and stunning power of flight. Yet not quite all—for in photos of condors in the wild I've come to appreciate their splendid red skin standing out against jet-black feathers, glowing in the sunlight. And gradually, their faces have grown on me, slightly comical, endearing.

At one time, California condors ranged widely—from Baja California in Mexico all the way up the West Coast to British Columbia in Canada—but by the 1940s, they had disappeared almost everywhere, except for an estimated 150 in the arid canyons of Southern California. In 1974, there were reports of two condors in Baja California, and my late husband, Hugo van Lawick, was asked to fly down to try to film them. But the expedition never materialized, and the birds disappeared.

The decline in condor numbers was due to many factors, such as the number of people moving into the western United States, shooting by poachers and collectors, feeding on poisoned baits set out for bears, wolves, and coyotes by ranchers, and, perhaps most importantly, the unintentional poisoning from lead ammunition fragments in the carcasses and gut piles of animals shot by hunters.

A group of biologists decided that something must be done. True, an area of wilderness had been set aside for the condors, but it was not enough. It served to protect them when they were nesting—and it was a preferred place for that—but when they foraged, they would fly a hundred miles or so into the ranchlands, where there was no protection at all. Noel Snyder, a biologist and passionate advocate for the birds, helped to establish the Condor Recovery Program and subsequently led the condor research effort. Biologists sought to discover all they could about condor behavior and the causes for the decline in numbers, while at the same time planning a captive breeding facility so that additional birds would become available to boost the wild population.

But there were many people vehemently opposed to any kind of intervention, and a controversy began that continued for years. The "protectionists" wanted to give the birds better protection in the wild, and if this did not work, to let them gradually disappear, to die with dignity in their natural habitat. They maintained that some condors were sure to be accidentally killed during capture; that they were unlikely to breed in captivity; and that, even if they did, it would be impossible to reintroduce them to the wild.

I remember visiting the San Diego Zoo during that period and discussing the issue with some of the scientists, including my longtime friend Dr. Donald Lindburg. Part of me shrank from the idea of depriving the wild birds of their freedom, imprisoning those wondrous winged beings in enclosures, perhaps for the rest of their lives. But another part felt—along with Don and Noel Snyder—that it would be worth it to save such a magnificent species, so long as they could be released back into the wild. In the end, Noel and Don and the other interventionists prevailed.

The Condor Becomes Extinct in the Wild

In June 1980, five scientists, led by Noel, set out to monitor the progress of the single chicks in each of the only two known "nests" in the wild. (For condors, nests are simply ledges of rock, usually in caves.) Imagine the team's dismay when, after they had checked on the first chick without problems, the second died of stress and heart failure during the process. This, naturally, led to a storm of protest from the protectionists—which Noel somehow weathered.

In 1982, a hide was built near a wild condor nest so that the behavior of the birds could be studied. The observers could hardly believe the extraordinarily dysfunctional behavior they witnessed. Every time the female returned to take her turn at incubating her egg, she was subject to violent aggression from her mate, who apparently did not want to relinquish care of the egg. The male repeatedly chased her from the nest cave, sometimes continuing to do so for days, while the egg meanwhile suffered unnaturally frequent and long periods of cooling. Finally, during one such squabble, the egg rolled out of the nest cave and smashed on the rocks below.

The observers thought that this meant a sad end to the pair's reproductive activities for the year. But a month and a half later, they produced another egg, which was laid in a different cave. Although this egg was also lost when the pair resumed squabbling—this time to a raven—the study was important because it established that condors, like many other birds, will be stimulated to breed again and lay replacement eggs if they lose one to predation or some kind of accident. Noel and his team then began a major effort to establish a captive breeding population by taking first-laid eggs from all wild pairs for artificial incubation.

How fortunate that they did, for over the winter of 1984–1985, tragedy struck the wild population. Four of the five known breeding pairs were lost. Reasons for the birds' disappearance were unclear, but there was mounting evidence that they were dying from lead poisoning. At this point, Noel and his team felt it imperative to capture the remaining wild birds. There were so few condors in the breeding program and they lacked the genetic variability to be self-sustaining—and there were but nine wild birds left. Only by establishing a viable captive population, Noel maintained, could the California condor be saved.

The National Audubon Society, however, strenuously opposed this plan, arguing that habitat could not be protected for the species unless some birds were left in the wild. In an attempt to prevent taking the last wild condors captive, the group sued the US Fish and Wildlife Service (USFWS). But after the female of the last breeding pair became a victim of lead poisoning and died, despite attempts by veterinarians to revive her, a federal court ruled that the USFWS did indeed have the right to capture the remaining wild birds. And so, between 1985 and 1987, the last wild California condors were taken into captivity, and the species became officially extinct in the wild.

Visiting the Breeding Center

By this time, two state-of-the-art breeding facilities had been established, one in the San Diego Wild Animal Park and the second in the Los Angeles Zoo, each with six enclosures. During five years, starting in 1982, sixteen eggs (of which fourteen hatched and survived) and four chicks were taken from the wild and shared between the two facilities. And there was one male, Topa-topa, who had been living in the Los Angeles Zoo since 1967. These captives were then joined by the last seven adults from the wild. Bill Toone was in charge of incubating the eggs, and thanks to the techniques he and his team developed, 80 percent of the eggs resulted in healthy young birds—compared with a 40 percent to 50 percent success rate in the wild.

In the early 1990s, Don invited me to visit the condor breeding center and flight cage at the San Diego facility. As with most such programs when reintroduction into the wild is an end goal, great care was taken to ensure that the captive-bred condors did not imprint upon their human keepers. Those caring for the chicks were equipped with glove puppets mimicking the head and neck of an adult condor, and no talking was allowed near the birds. In silence I peered through one-way glass and saw one of the original wild hatched females sitting, unaware of my presence, on a ledge of man-made rock. As I watched her suddenly take off and, with only a couple of flapping movements, glide on those majestic wings across her very large flight cage, I felt tears sting my eyes. Partly because of her lost freedom; partly because I knew that but for a handful of passionate, courageous, and determined people, this glorious winged being would almost certainly have died—shot or poisoned—like so many others before her.

More than two decades later, in April 2007 (on my birthday!), I visited the Los Angeles breeding program and met team members Mike Clark, Jennifer Fuller, Chandra David, Debbie Ciani, and Susie Kasielke. We gathered in a small room where video screens showed the twenty-four-hour recordings of behavior in the breeding enclosures. As we talked about the successes and problems of the program, we watched on a monitor (from a remote camera set up in a breeding pen) a wonderful courtship display by a young male. And there was a female there that was getting ready to lay her first egg. It was not due for several days, but already she was looking most uncomfortable, her tail raised and her head low. She pecked up and swallowed a few small fragments of bone, a behavior thought to provide extra calcium for building the egg's shell.

Preparing Young Condors for a Life in the Wild

An urgent problem facing the pioneers who embarked on captive rearing was to find the correct method of raising the young birds for ultimate release. Because the California condors were so perilously close to extinction, they could not afford to make many mistakes. So the team decided to carry out a trial release with Andean condors, since this species, with its fabulous eleven-foot wingspan, is not nearly so endangered. Thirteen youngsters would be raised and released temporarily in Southern California, allowing the team to test their methodology before any of the precious Californians were freed. The Andeans, all females, were raised as a peer group and all released at the same time. It was thought that they would provide one another with companionship and support one another. And indeed, it worked well. (The Andean condors were later recaptured and ultimately re-released in Colombia, where many are now breeding and rearing chicks of their own.)

Flushed with the success of the Andean program, the biologists confidently raised their young California condors in the same way. Alas, Mike told me, group rearing simply did not work for them and led to all manner of behavioral problems. It seems that California condors need discipline from an adult bird. And so a new method was devised. Each chick remains for the first six months in a solitary nest box, in view of an adult male condor, cared for and fed by a disguised human using a condor head puppet. Then, at the time when a wild fledgling would leave the nest, the youngster joins an adult mentor—a male of ten years or older. This mentor competes for food with the young bird but without being aggressive and is, said Mike, "good for its mental development."

However, as the condors matured, additional behavioral problems arose. For one thing, proper male–female bonding was not happening until, through trial and error, scientists learned that putting a mature male and female, genetically suitable for each other, in an enclosure with young birds worked best. "Each adult bird then prefers the other's company over that of any of the youngsters," said Mike.

Once bonding is achieved, mating is no problem, and such pairs regularly produce eggs. And the raising of chicks by parents in captivity was also relatively trouble-free. "The sight of an egg," said Mike, "seems to trigger an instant paternal response in the male, who becomes very protective of it." The pair take turns incubating the egg for the fifty-seven days before it hatches. After this the male continues to be very protective, though the mother tends to compete with her chick for its father's attention.

Odd Parenting Back in the Wild

By 1991, eleven of twelve captive pairs had produced twenty-two eggs. Seventeen of these were fertile, and thirteen had hatched and matured. Things were going well. By 1992—less than ten years after the program began—the first two captive-bred condors, each with a radio tag, were released into 398,000 acres of protected wilderness, including thirty miles of protected streams, in Los Padres National Forest. In an attempt to protect these birds as much as possible from the risk of lead poisoning, food was (and still is) set out near the release site. Even though they can fly more than a hundred miles in a single flight, it was hoped that these California condors would, as had the test group of Andean condors, return to easily available food when hungry—which they mostly did.

In 2000, the first captive-bred birds nested in the wild—an event that is always awaited eagerly by the people who have worked so hard to return animals to a life of freedom. But it was at this time that some of the behavioral problems affecting the captive-raised birds became apparent. When biologists found the nest, they were amazed to see not one but two eggs! And they discovered that there were three birds to this one nest, one male and two females. They had, however, chosen a very appropriate cave, where the females had laid eggs several feet apart. The three took turns sitting at the nest site—but one bird could not sit on both at once, and so the biologists decided to intervene.

The first condor to be provided with a radio transmitter was IC1 (shown here with Noel Snyder, left, and Pete Bloom), who was trapped in the Tehachapi Mountains. Transmitters make it possible to track the progress of condors in the wild. *(Helen Snyder, courtesy US Fish and Wildlife Service)*

They found that one of the eggs was completely rotten. They then left a dummy egg and took the other to see if it was viable. It turned out to be in poor shape, but the skilled staff managed to hatch it at the zoo. Meanwhile, the unlikely trio was still caring for the dummy egg in the wild. Just before the egg should have hatched, it was replaced with a healthy, captive-laid egg. A chick duly hatched, but despite the presence of three potential caregivers, one of the females was left alone—first with the egg and then with the chick—for eleven days straight. And when the second female finally returned, instead of helping to nurture the three-day-old chick—she killed it. That was certainly not a very successful breeding season! Still, it was encouraging that the three would-be parents had nested in a suitable location and among them had at least hatched an egg.

Trash and Other Troubles

The following year, chicks were hatched in three nests. But initial excitement turned to dismay when, at about four months old, all three youngsters died. When they were subsequently examined, it was found that the parents, in addition to providing them with normal food, had been feeding them trash—items such as bottle tops, small pieces of hard plastic and glass, and so on.

Unfortunately, this has become a tradition in this population—and they are not alone, as vultures in Africa have also been observed feeding trash to their young. Biologists believe that the parents are picking up these inappropriate objects as substitutes for the bone fragments thought to help in bone development.

Today the recovery team keeps close watch on the nests to record parental behavior and chick development, and they check the health of eggs and chicks at regular thirty-day intervals—with a mandate to intervene if necessary. And it was necessary with the only chick hatched in the 2006 breeding season. This is a fascinating story. For one thing, the parents had both been considered too young to produce an egg—the female was only six years old and the male, only five. They had not even acquired adult plumage, and finding them with a nest and egg was a huge surprise. Mike told me that they were all worried because of the youth and inexperience of the parents—would they be able to sustain interest in the egg during the long incubation period?

So the team played a trick. Members took away the egg of the inexperienced couple, which would not hatch for another month, and left in its place an egg from the captive breeding program that was on the verge of hatching. The young parents, hearing the vocalizations of the chick inside the new egg and the pecking at the inside of the shell, instantly became very attentive. The chick hatched successfully and was well cared for.

When its health was monitored after thirty days, all seemed to be going well, although some trash fragments had been brought to the cave floor. The team spread five pounds of bone fragments around, hoping this might mitigate the extraordinary passion for feeding trash, and left, hoping for the best. The sixty-day checkup also found the chick healthy. The parents had left more trash fragments lying around, but the metal detector—now standard veterinary equipment!—showed that the chick had not swallowed any. However, when they checked up after ninety days, they found a very sick, underweight, and undersize chick that had swallowed a great deal of trash. It was obvious he would die if this was not removed.

Mike picked up the chick and took it back to the Los Angeles Zoo—which does veterinary work on all the wild condors in California—for emergency surgery. Meanwhile, another member of the team stayed overnight to keep the parents out of the nest site—for had they found it empty, they would almost certainly have left. Inside that chick was an extraordinary array of trash, ranging from bottle caps to small pieces of metal and hard plastic, all tangled in cow hair. I saw the collection, and it was hard to believe it all came from one bird, let alone a chick. No wonder it was sick! The surgery went well, and twenty hours later the youngster was returned to its nest by helicopter and delivered on the end of a rope by a search-and-rescue specialist. During this operation the parents were right behind the humans, peering past them at the nest—and five minutes after the helicopter left, they were back with their beloved offspring.

Without his load of indigestible trash, the chick's health improved. But just before the 120-day check, the field biologist on duty, observing the nest through a high-powered scope, noticed the chick playing with three pieces of glass, swallowing them and spitting them out. And sure enough, when team members went in to check him at the prescribed time, they could feel something hard in his crop. Fortunately, they were able to massage the objects gently out of the crop and into the throat, then remove them with forceps—they were the same three pieces of glass that he had been seen playing with. This preoccupation with trash is certainly one of the worst behavioral problems that the team must try to solve.

One suggestion to reduce behavioral problems was to release some of the original wild-caught birds from the 1980s to serve as role models. This was done, but while these birds do indeed represent a priceless behavioral resource, part of their behavior is wide-ranging foraging, which can make them especially susceptible to lead poisoning—and in fact, one of the original females did suffer serious lead poisoning after her return to the wild. Noel feels strongly that no more should be released until the lead-contamination problem has been solved.

Faith in the Future

From the very start, Noel told me, nearly all program personnel have agreed that this issue is critical. But for more than twenty years—since the first sick condors were diagnosed with lead poisoning—nothing was done to remove the source of the problem, largely because no good substitutes for lead bullets existed. By 2007, however, a variety of nontoxic ammunition had come on the market, and on October 13 that year Bill AB 821 prohibiting the use of lead bullets for hunting large game in the range of the California condor was signed by California Governor Arnold Schwarzenegger and subsequently passed by the legislature. This was the result of pressure on the lawmakers from the Natural Resources Defense Council (NRDC) and many conservationists.

Some environmentalists felt that the bill was a cop-out—that so long as the bullets were made, it would be hard to enforce the law. But when I talked with Governor Schwarzenegger about this, he said that the range of the condors was so vast, there was not much of California left where lead bullets could be used. He thought manufacturers would not think it worthwhile to continue making them. In any event, passing this bill is a major step forward, and I, for one, congratulate the governor for supporting it.

Although the future of the released individuals is not assured, the investment, in time and the commitment and dedication of the men and women involved, has been a success—for without intervention, the California condor would most certainly have gone extinct. Instead, there are nearly 300 of these magnificent birds, and 146 of them are out in the wild, soaring the skies above Southern California, the Grand Canyon region of Arizona, Utah, and Baja California.

Those who watch the condors in the wild are moved. Mike Wallace, one of the field biologists who oversaw the release of captive-bred condors in the Baja, sent me a wonderful story about observing the mating rituals and unique personalities of these amazingly social birds (which you'll find on our Web site janegoodallhopeforanimals.com). My friend Bill Woolam wrote to me about the wonder of seeing this giant bird when he was hiking in the Grand Canyon—watching the condor flying up and up with those huge and powerful wings, hearing the wings flapping and the air whistling through the feathers as the condor glides down—the music of flight. And Thane, too, recently wrote to me about the joy of seeing five of the fifty or so condors living near the Grand Canyon when he was rafting there in 2008.

The more people who have this kind of experience, and who realize how nearly this amazing bird vanished forever, the more they will care. And their number is growing—there are legions of people who are passionate about California condors and their future. Noel, though officially retired, still feels a great personal commitment. The condor, he told me, "comes to dominate your life whether you like it or not."

I have a legal permit to carry a twenty-six-inch-long wing feather from a condor. During my lectures, as Thane mentioned in his foreword, I love to take this by the quill and pull it, very slowly, from its cardboard tube. It is one of my symbols of hope and never fails to produce an amazed gasp from the audience. And, I think, a sense of reverence.

Archival photo. One of the original milu (Père David's deer) herds relocated to Woburn Abbey. After the milu became extinct in China in 1900, the eleventh duke of Bedford gathered together the few remaining milu from European zoos and brought them to Woburn Abbey. Thanks to his foresight, the milu was likely saved from global extinction. *(By kind permission of the Duke of Bedford and the Trustees of the Bedford Estate)*

Milu or Père David's Deer
(*Elaphurus davidianus*)

The first time I was able to see this rare and beautiful deer in its native homeland was in 1994, during my first visit to China. Dr. Guo Geng showed me round the Nan Haizi Milu Deer Park, just outside Beijing. Guo Geng is enthusiastic and passionate about his work, which includes the education outreach for this park. A small part of the park was like a zoo—enclosures held various deer and a few other hoofed animals—but there was also a large fenced-in wilderness area, complete with small lake, home to a herd of Père David's deer—known in China as milu. How magnificent they looked grazing near the shore of the lake. They wore their grayish brown winter coats—but, said Guo Geng, their color changes to reddish brown during the summer. They are similar in size to the red deer of Scotland. One handsome male stood, seeming to look directly at me, proud and dignified. I could see no fences, no boundary to his wild space.

As I stood there watching the milu, my mind suddenly jumped far back in time. I vividly remembered visiting a herd of these deer on the duke of Bedford's estate in England, and hearing that they were highly endangered and had originally come from China. That was in 1956, when I was working with a documentary film company in London and we were making a film at the estate. And now, forty years later, I was looking at some of the progeny of those very deer.

Extinction in China

Their story amazes me. The milu was once common in the open plains and marshes along China's lower Yangtze River Basin. But, mainly as a result of habitat loss and probably some hunting, they were on the brink of extinction by 1900. The last known wild individual was shot in 1939 near the Yellow Sea. Fortunately for the survival of the species, the emperor of China had installed a large herd in his Imperial Hunting Park (Nan Haizi Park) near Beijing. The deer thrived in this park, which was surrounded by a forty-three-mile-long wall and guarded by a Tartar patrol.

In 1865, Père Armand David, a French Jesuit missionary, introduced the deer to the Western world. He had been passionately interested in nature from childhood, and had always wanted to go to China. He became a missionary, and his dream was realized when he was given a leave of absence for five months to tour in China. During this time, he collected numerous undescribed (at least to Westerners) plants and insects and sent them back to the natural history museum in Paris for study. He also described the golden monkey, some pheasants, and a squirrel, and was the first person to describe a giant panda to the West.

During one of his travels, just outside Beijing, he came upon the wall that concealed the Imperial Hunting Park. Managing to look over, he saw some strange animals that looked a bit like reindeer, but he soon realized they were not. Back in Beijing, he tried to discover something about them and, failing, returned with an interpreter to the hunting park. Eventually, by providing

Historical photo of Père David himself, an extraordinary naturalist and explorer: savior of the milu. *(Reprinted with the permission of the Deandreis-Rosati Memorial Archives, DePaul University, Chicago, IL)*

some woolly caps and mittens to the guards (though one story says twenty silver pieces), he persuaded them to bring him some pieces of antlers and skin. Père David sent these precious specimens back to France, where they were examined and pronounced to be from a new species of deer—which was named in his honor.

There was a keen desire in Paris to obtain some live specimens. Eventually, after many failed attempts, the Chinese emperor was persuaded to gift three individuals to the French ambassador. Sadly they did not survive the arduous sea journey. But after further negotiations with the imperial staff, a few more pairs of the deer were gifted and this time arrived safely in Paris. There was much excitement about the arrival of the first Père David's deer; eventually zoos in Germany and Belgium, as well as the Woburn Abbey park in England, also acquired specimens.

Soon there were approximately two dozen deer in Europe, in addition to the large herd remaining in China, and the survival of the species seemed assured. But in 1895, catastrophic floods devastated China, and a part of the wall surrounding the imperial park was destroyed. Many of the deer were killed by the floods; others that escaped through the breach in the wall were hunted and killed by the starving population. Still, between twenty and thirty survived in the park—enough to maintain the species. Alas, five years later they perished during the Boxer Rebellion, when troops occupied the imperial park and killed and ate every single deer.

Surviving in Europe

Thus the future of the deer depended on the few individuals in Europe—and the zoos found that they were reluctant to breed. When news of the slaughter of the last deer in China reached Herbrand, eleventh duke of Bedford, he realized the need to consolidate the scattered groups if the species was to be saved. Eventually he persuaded the various zoos to sell their animals, and by 1901 he had collected a total of fourteen Père David's deer in the park at Woburn Abbey—the last individuals in existence. There were seven females (two of whom were barren), five males (one of whom established himself as the dominant stag), and two youngsters. It required years of patient management before these last survivors of a once abundant species began to breed.

In 1918, when the population numbered around ninety animals, they suffered yet another major setback: World War I caused widespread food shortages in Britain, which meant not enough food for the exotic deer, and the population was reduced to just fifty. After the war, numbers again began to increase, but in 1946, when the population of Père

David's deer had risen to three hundred, World War II created more shortages of food—and in addition, the herds were threatened by nearby enemy bombing. At this point, the duke of Bedford realized it would be wise to spread out the breeding population. By 1970, there were breeding groups of Père David's deer in centers all over the world, with over five hundred at Woburn Abbey alone.

Planning the Return of the Milu

The decision to try to reintroduce these Chinese deer to their homeland was the idea of the then marquis of Tavistock, later the fourteenth duke of Bedford. It was not an easy operation, but finally, in 1985, twenty-two Père David's deer—which would henceforth be known as milu—set off from Woburn Abbey to Beijing, accompanied by one of their keepers. In 2006, during my annual visit to Beijing, I told Guo Geng that I needed to know more of the history of the deer's return to China. He told me that I should talk to a Slovakian woman, Maja Boyd. We planned to meet in Beijing, but sadly that meeting never took place as her cousin died, suddenly, and she had to fly back to Slovakia. However, just before Christmas that year, we spoke by telephone—she was in Slovakia, and I was in Bournemouth.

By the end of the conversation, I felt I had tapped into Maja's warm and giving personality. She wanted me to know that when her late husband had first taken her to America, she had watched a film about me and the Gombe chimpanzees. "I so badly wanted to do something like you!" she said. Her American husband had been a good friend of Lord Tavistock, as the duke of Bedford then was. And when Maja learned about his plan to send Père David's deer to China, she was fascinated. "It was the deer," she told me, "that took me to China."

She would have loved to release the deer into a really wild place. "But," she said, "the government chose the site, and we needed their full support." And it made good sense, for the place chosen for the deer park was once part of the Imperial Hunting Park as well as being close to the center of government in Beijing.

Maja had gone to inspect the area prior to the return of the deer. She found that part of it was a tree nursery—which was fine. But there was also a pig farm, which Maja felt was not appropriate. The government agreed to move the pigs. Then they had to block access for a stream that flowed through the area, since it was horribly polluted. They dug nine

Maja Boyd, a guardian of Père David's deer, shown here with a hand-reared female deer at the Nan Haizi Milu Reserve, Beijing. The young deer's mother died soon after giving birth. Maja told me that this deer "followed me around like a dog." *(Maja Boyd)*

little wells to provide water for the animals and embarked on the major project: filling the lake with clean water.

The new arrivals deserved the best the Chinese could give them. But there was another major problem. The officials in charge of building the required quarantine sheds insisted that they be designed like the traditional stall for cows or horses—with a half door. No matter how often Maja explained that deer were different, and would immediately leap over a half door, the Chinese would not, or could not, believe her. Matters came to a head when Lord Tavistock's eldest son, Andrew Howland, arrived to inspect the accommodations for his precious deer. He was horrified when he saw the row of sheds with half doors, and insisted that they physically break down the doors. After this the doors were rebuilt—correctly! Finally, all was ready.

Return to the Ancestral Homeland

And so, in 1986, the twenty-two deer that had been born on an estate in faraway England—some of them, perhaps, offspring of those I had seen when I visited Woburn Abbey in 1956—set off for China. It was a long plane journey but much quicker than the sea voyages their ancestors had endured. Maja vividly remembers the day they arrived. She found it fascinating that they were traveling Air France. "These deer were first introduced to the Western world by a French missionary, and they came back on a French plane." Everyone was so excited that they forgot what they were supposed to be doing as they struggled to get closer for a first glimpse of the historic cargo. The containers were jostled, and both Maja and the keeper who had traveled with them from the UK feared the cages would fall and the deer escape. Fortunately, although they had not been sedated, the deer themselves remained very calm. "In fact," said Maja, "they behaved much better than the humans present!"

Finally all the cages were loaded onto trucks, and the deer set off on the last part of their long journey. Maja said she felt so sorry for the hundreds of excited people who lined the roads, hoping for a glimpse of the new arrivals, because all they saw were the trucks. What a moment when the deer finally entered their quarantine quarters and stood on Chinese soil—where their ancestors had roamed half a century before. Right from the start the Chinese were very proud of the project, and there was a great deal of publicity. Children, in particular, were interested.

"We got a lot of letters from kids," Maja told me. She remembered one in particular from a five-year-old girl. Her parents had given her two RMB (at the time, this would have been about seventy-five cents) for a month's pocket money. She sent it to the deer park and asked if they would "please buy chocolate for the uncle and auntie Milu so they know they've arrived in a country that welcomes them."

There was one unexpected outcome of the return of the milu. When local villagers heard about the deer park, they realized that it would be a perfect place, quiet and green, for burying the cremation remains of their loved ones. And so, after a death, they went there and dug little graves in the park. Maja told me she was once walking the grounds with a Chinese official. He looked at the graves and announced, "We must eliminate these." But Maja told him that in her native country—Slovakia—it is very bad luck to desecrate a grave. The official looked around, took her by the hand, and whispered that they, too, feel the same. So today there is a special place where one can see little mounds—and the people have permission to return there every year during the Qing Ming Festival at the beginning of April, when Chinese pay their respects to the dead.

Visiting the Père David's Deer at Woburn Abbey

Maja arranged for a few of the Chinese scientists involved with the Père David's deer to visit the UK, and a highlight was their visit to Woburn Abbey. There they would meet the people who are working to maintain the herds outside China. I was hoping to join them, but unfortunately the Chinese delegation arrived the day I had to leave for America. Still, I was able to meet Maja for the first time during my visit to Woburn Abbey, and Lord Robin Russell (son of the duke of Bedford) was a charming host.

For almost a week it had been raining, but after my sister Judy and I had driven all day in heavy rain, the sun came out to create a glorious spring evening. The grass was brilliant green, the old oaks a softer olive shade. At first, the only Père David's we found was one that had "double-shed"—lost his antlers before the rut and not yet grown new ones. Without them he could not compete with the others, and was probably wise to avoid the herd. We passed herds of sika deer, roe deer, fallow deer—and the spectacular red deer. Where were the Père David's? We searched and searched, and finally found them down where it was very wet. What a wonderful sight—about two hundred of them, their coats a rich golden color in the light of the setting sun.

Too soon twilight began to fall and we had to leave them. But then, in the charming old cottage where Robin lives with his wife, we sat and talked deer. I got to know Maja better and learned more about the history of the Père David's project. Robin generously offered me access to the photo archives. And we discussed forming a collaboration between their education program and the Jane Goodall Institute's Roots & Shoots.

A Final Dispatch from China

During my Asia tour in the fall of 2007, Maja arranged for me to revisit the milu park outside Beijing. There I was very pleased to meet two of the delegation to Woburn Abbey whom I had missed in the summer:

Director Zhang Li Yuan, and Chinese Professor Wang Zongyi, who has been so instrumental in reintroducing the deer and such a very big help to Maja. After sitting and talking (with Maja translating) and drinking hot tea, we set off on a golf cart to see the deer. It was bitterly cold with icicles hanging down from some of the trees, and I was glad I had dressed warmly.

That tour depressed me. The first time I had visited the park, there had been a real feeling of being in the countryside, even though it is so close to Beijing. But now there is development pressing in from all sides. The herd of milu had grown. They had eaten all the available grass so that, especially in the winter, they needed supplementary food. They appeared healthy enough, but they were standing around their feeding troughs looking somehow weary—bored, perhaps. They almost looked like a different species from what I had seen in 1994; the sense of freedom and nobility that had been so strong during my previous visit to this place was no longer there.

We were glad to get back inside the comparative warmth of the little environment center. As we enjoyed a truly delicious vegetarian lunch, my hosts told me about the twenty-five-hundred-acre nature reserve in Shishou in central China, on the Yangtze River. At the beginning of the 1990s, I heard, China's National Environment Protection Agency had agreed that a small herd could be moved to this area, where they settled down well. And some individuals swam across the river and started a truly free-ranging population on the other side, in Hunan province. At first there were concerns that they would be hunted, but instead the local population reveres and protects them. Both Maja and Professor Wang Zongyi begged me to make time to go and see these milu, living in the wild as they did so long ago, and one day I should love to do so.

In the meantime, I carry around a glass medallion, given to me by Guo Geng, embossed with a drawing of the milu made during the Han dynasty (206 BC–AD 220). And at our JGI office in Beijing, we have an antler, shed by a four-year-old stag, that I take to lectures when I am in China as one of my symbols of hope. It represents the resilience of animals if we just give them a chance. Since returning to China in 1985, the milu have prospered and their numbers have increased. There are about a thousand now, all told.

Red Wolf
(Canis rufus)

When I was a child, I loved the legend of Romulus and Remus, the twins raised by a she-wolf in the forests of Italy. It gave a strange sense of authenticity to my favorite wolf story of all—the adoption of little Mowgli into a wolf pack, in Rudyard Kipling's *Jungle Book*. And then came Jack London's *Call of the Wild*, which not only reinforced my love of the wolf, but gave me a passionate longing to spend time in the wilderness with these magnificent animals.

It is unfortunate that wolves have been so hated and so feared. There are very few authenticated accounts of wolves attacking human beings in North America. Occasionally they will take livestock because, of course, we have moved farther and farther into their wild hunting grounds. And because of this, along with fear, they have been horribly persecuted in Canada, the United States, and Mexico—trapped and poisoned and hunted with bows and arrows, spears and guns. Even attacked from the air by people in helicopters. And in light of what we now know, thanks to numerous wildlife biologists who have spent years observing them in the wild, the all-out attempt to eradicate wolves can be seen as tragic, unjustified—and in a way extraordinary since they are indisputably the ancestors of "man's best friend," the domesticated dog.

There are three species of wolves in North America, of which the gray wolf is the best known. Then there is its close cousin, the Mexican gray wolf. And the red wolf, the subject of this chapter. The three species have many similarities in behavior. A pack typically comprises a breeding pair and their offspring—yearling pups from a previous litter and the pups of the season. They are most active in the early morning

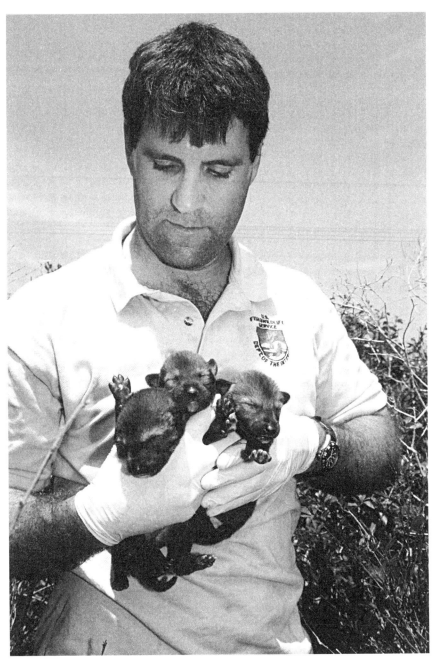

Art Beyer, USFWS wildlife biologist, checking out the health of wild pups, just a few days old. The parents will come back after the biologists leave and move the pups to a different secret location. *(Melissa McGaw)*

and evening when they hunt as a pack. Small cubs, of course, stay in the den—initially with their mother—and other pack members return to feed them by regurgitating meat.

Red wolves are recognizably smaller than gray wolves and about twice as large as coyotes—although yearling red wolves are almost the same in size and coloring as adult coyotes. At one time they were common throughout the southeastern United States, but predator control along with loss of habitat severely decimated their numbers during the 1960s, until only a few remained along the Gulf Coast of Texas and Louisiana.

By 1973, when the red wolf was finally classified as endangered, it was on the very brink of extinction. Scientists decided, in a desperate bid to save the species, to capture as many as possible for captive breeding with the goal of eventually returning them to the wild. Only seventeen were found. When the last of these was captured in 1980, the red wolf was declared extinct in the wild. All red wolves in existence today are descendants of fourteen of those individuals captured in the early 1970s.

From Pen to Freedom

The breeding program, in which a number of zoos took part, was coordinated by the US Fish and Wildlife Service's Red Wolf Recovery Program. By 1986, it was thought that there were enough young captive-born wolves to start the release program, and after careful surveys North Carolina's Alligator River National Wildlife Refuge had been selected as the most suitable area. And so, fourteen years after the birth of the first captive litter of red wolves, four adult pairs were taken to their new home.

Of course, not everyone was thrilled at the idea of wolves roaming in the wild again. And so, in order to convince the public that, if things went wrong, the wolves could be easily recaptured at any time, scientists had been working on collars that could be remotely activated to discharge an anesthetic into the animal concerned. Unfortunately, these were not ready in time, and the four wolves, and others who followed, had to be held in large fenced enclosures for almost a year, much longer than planned. At least they had time to get accustomed to their new environment—its scents and sounds and some of the various animals they would meet in the field. And finally the day came when the first pair of wolves could be released to begin exploring their new wilderness home. The other pairs were released at weekly intervals.

It was a heady time for the Red Wolf Recovery Program field team. Chris Lucash, who continues to devote his life to this program, was part of the original team. I asked him how he felt when the wolves were first released. "How I felt? Wow! Excited, elated, incredibly—and naively—optimistic. I felt extremely fortunate, maybe even blessed, to be in a place at a point in time that was so rare and potentially such a pivot point in history, at least for one very historically unlucky species. This was *the* most important thing I could be doing." It was a time, he told me, when they were filled with hope at every release.

They did not fully realize the dangers these naive wolves would face. They did not guess that 60 to 80 percent of those wolves would not make it—would get sick or collide with a car as they tried to cross the roads that bisected their new home. And the field team felt devastated by each loss. "We had to learn to keep some distance, try not to get too emotionally involved," said Chris. That was one of the reasons why the wolves, for the most part, were not given names.

But it was impossible to remain absolutely detached, especially back then. There were only a few wolves, and the biologists knew them all personally; they handled them, followed their movements, tried to understand their behavior and motives. And when they had to capture them, they had to find ways to outwit them. "Our hopes and spirits rose high with the good news, and sank deep with the bad. Those of us here from early on had to do a lot of growing up—and it just happened to be in association with these animals." Chris and Michael Morse, another biologist from those first days who is still with the team, have shared some of those early stories with me.

A True Survivor

Although the wolves are not officially named, for convenience the field team gave them names that usually derived from the location of the pack or some nearby geographic feature. "Not too romantic, but better than a stud book number," said Chris. And, for the most part, those are the names that I have used. *Survivor*, though, is the name that I have chosen, retrospectively, for the first wild-born pup of the recovery program—because she survived against incredible odds.

"Her captive-born parents were physically impressive and beautiful, but ill-fated," said Chris. It seems they only had the one pup; the biologists did not disturb the den at the time, but looked for signs later. A few weeks after whelping, Survivor's mother crawled back to the release pen and into the den box from which she had been released about eight months before. And there she died of a uterine infection. Survivor, who could barely have been weaned before her mother died, survived, presumably, with the help of her father. Alas, a few months later she lost him, too—he died of asphyxiation with a raccoon's kidney lodged in his trachea. For weeks, then months, there was no sighting of Survivor, although sometimes the team found tracks that could have been hers. And, indeed, against all odds she had survived.

Eventually she was captured and collared. She managed to evade death during a private trapping season (when trappers are allowed to trap furbearing animals as a form of nuisance control or as a "hobby"). In fact, she became very smart at avoiding traps, and when the team wanted to catch her—in order to replace her collar, for example—they had to work hard to outsmart her.

She eventually paired with a male, and they became the first wolves allowed to stay on private land, south of the refuge. After this, Survivor was captured—yet again—to replace her collar. It was to be for the last time, for the new collar stopped working, and they never found her again.

Brindled Hope

Brindled Hope was one of the first wolves to be released in late 1987. It was months after her arrival from a wolf sanctuary in Missouri that they noticed her name, handwritten in small letters, on the back of her skykennel. She was not a very impressive-looking wolf, Michael told me. She was smaller than average—and at five years, older than most selected for release. Nevertheless she and the mate who had been chosen for her produced one of the first two pups born into the wild that year. The pup was a female, officially 351F but whom I am calling here Hope.

It was not long before disaster struck: Brindled Hope's mate was killed by a car on the highway when their pup was only a month old. Brindled Hope, not knowing, waited for him as long as she could, but she needed to move to an area where there was more prey. And so, after eleven days, she set off toward the more open farmlands where she had once hunted with her mate. She and her pup traveled beside the highway, quickly moving into the thick vegetation whenever a car approached. There the team found the two of them, the pup struggling to keep up with her mother. Keeping their distance, the biologists followed until they reached a dirt road leading to the safety of the fields. First, though, the mother and pup had to cross the highway—and the biologists stopped traffic in both directions until the pair made it across. Brindled Hope successfully raised her pup, Hope, and eventually mother and daughter paired with the Bulls Boys and lived in their pack for many years.

The Bulls Boys

Biologists prepare some captive wolves for release by raising them in wild settings on islands within wildlife refuges, where they can learn the survival skills they will need for their new life. Such were the Bulls Boys, brothers who arrived as yearlings in 1989 after living for almost a year on Bulls Island at Cape Romain National Wildlife Refuge in South Carolina. They were released into what was known as the Milltail Farms area on the Alligator River National Wildlife Refuge. "We had no clue that they would catapult the fledgling wolf project on the road to success," said Michael. "With their tall, lanky bodies, sizable feet, and broad heads, their appearance, although impressive, gave no hint of the substantial impact they would have on the recovery program."

The Milltail Farms area comprised some ten thousand acres of farms and forest where Brindled Hope had lived with her pup, Hope. When Hope was old enough to get by without her mother, Brindled Hope was recaptured, was paired with a new mate, and produced four new pups in captivity. Then she and her new family—including her mate—were released back into the Milltail Farms area. Surely, thought the biologists, there was plenty of space for all. But the Bulls Boys—the Milltail pack—were not pleased, and within a month had attacked and killed the male intruder. Soon after this, one of the brothers—I'll call him Boy One—paired with Hope; the other, Boy Two, paired with Brindled Hope, whose four pups, remarkably, were allowed to remain unchallenged.

It seemed that the Bulls Boys might each sire a litter in the next breeding season, and excitement ran high in the field team. "Second-generation pups were a major measure of the recovery program's success, and it was happening in the first two years!" said Michael. But as he said to me, "It was all too good to be true." Boy One, from Bulls Island and not familiar with roads, was killed crossing a highway just before the 1989 breeding season.

However, the surviving brother grew stronger and stronger. In 2000, he reached the advanced age of twelve years and, no longer a "boy," became "the Old Man." He actually allowed one of his sons to establish and raise a family "virtually next door," in part of his territory. This was an arrangement that he probably would not have tolerated in his younger days, speculated Michael.

"But even though the Old Man may not have been the breeding male of his pack in his last days," Michael wrote in a letter to me, "he left a living legacy." By the time he died in 2002, he had sired at least twenty-two pups from seven litters. "His genes are today an integral part of the wild population of red wolves in northeastern North Carolina." Reading between the lines, I sensed that Michael had a deep affection for this wolf. And I knew I was right when I came to his last line: "And I hope it's true, what the old-timers say—'All dogs go to heaven.'" For what it's worth, Michael, I'm sure they do.

The Gator Pack

The wolves from Graham, Washington, here called Graham Male and Graham Female, ultimately became the breeding pair of the Gator Pack. They had arrived together at the start of 1988 and were released with mates who had been chosen for them. Those matchmaking efforts, however, were not successful: The two females that were successively offered to Graham Male were killed by cars, and Graham Female's mate simply disappeared. And then Graham Female and Graham Male found each other and began consorting in the winter of 1989. They soon became inseparable: "Once they bonded, they were rarely apart," said Michael. Both grew to be very large in their prime, the male weighing a record eighty-four pounds and the female, sixty-five pounds.

Their home range was a vast sixty thousand acres of gum swamp and pocosins in the central portion of the Alligator River National Wildlife Refuge—a relatively harsh environment compared with Milltail Farms. "Seldom seen by humans," wrote Michael, "the Graham pair—now the Gator Pack—lived in near seclusion," producing three litters. In 1992, a family group of wolves was released near one of the boundaries of the Graham Pair's territory. "They ran off the adult pair and killed and ate the pups," said Michael. "It was the last time we attempted a release near them."

On April 1, 1994, Graham Male, aged nine years, was found dead in his territory: "It looked like he just lay down and died," Michael noted.

Red Wolf Recovery Team biologists Chris Lucash and Michael Morse check a litter of wild pups in northeastern North Carolina. The biologists perform a general health evaluation and insert a small transponder for identification purposes. *(USFWS)*

And only four months later, his mate left the Gator Pack territory and went on "a long walk-about." As she was passing through the home range of another wolf pack, the River Pack, to the north, "she lay down at Deep Bay to die."

Fostering Pups in the Wild

And so, gradually, the captive-born wolves adapted to their wild home, gave birth, and raised pups. Despite the heartaches and disappointments, there were many success stories, too. The team became more confident as their understanding grew regarding what could and what would not work.

Even when it became clear that the reintroduction program was a success, it was still necessary (and remains so to the present) to maintain the captive population at about two hundred individuals. This is in part because additional wolves are needed to bolster the wild population, serving as a backup in case a disease wipes out those in the wild, and in part to provide stock for future reintroduction programs in different areas.

A few captive-born pups are returned to the wild very early in life—when they are between ten and fourteen days old, just before their eyes open. At this age, they are readily accepted and cared for by both the male and the female of a wild pack. This "fostering" is only done if a wild mother has lost all or some of her own litter, or has a very small litter that allows her to manage one or two extra pups. Fostering of this kind not only boosts numbers of wolves in the wild but also, because pups are selected carefully, helps maintain the genetic viability of the population. I was fascinated to hear about this, and how it all began.

The first time it was tried was in 1998. "It was," Chris told me, "a bit of an act of desperation and/or lack of alternatives." A captive female killed one of her three newborn pups, and when it was found that she had done this before, at the small zoo where she had been kept, it was decided not to take chances with the two remaining pups. They were taken from her and, rather than being hand-raised, were placed in the den of a wild female. The team believed this would work based on experience with captive fostering, but nevertheless it must have been a wonderful moment when those little pups were immediately accepted by the female, who raised them with her own youngsters.

Sometimes the field team comes upon wild pups that have to be fostered. Once, a female was discovered dead in the area where it was believed that she had a den. A search revealed two pups, weak and dehydrated but still alive. They had been without their mother for at least two to three days. "After two days of reviving them as best we could," said Chris, "we located another wild female with similar-aged pups, who accepted and raised the fostered ones as her own."

Collars and Radio Tracking

Approximately 65 to 70 percent of the wild red wolves in northeastern North Carolina are wearing telemetry collars, either the standard VHS variety or one of the new specially designed GPS-enabled collars that use satellites to automatically record their location—and that of the wolves wearing them—four or five times each day. This information is stored in each collar, and every one to two months the biologists can download it all at once with a special receiver. These data—which can consist of three to four hundred locations!—are then used to create a map that will show movement patterns, habitat preference, and home-range size as well as proximity to any other wolf who happens to be wearing a collar.

Michael sent me, from one of his reports, an example of wolf tracking with this technology. Wolf "11301M" was collared as a yearling when he was still living with the pack in his natal home range. Over the next year, the data regularly obtained from his collar provided the field team with a wealth of information. First they learned about his movements in his original home range. Then, when he left his natal area in the spring and began his travels, they learned where he went.

"He seemed to go from wolf pack to wolf pack looking for a place to live," wrote Michael. ". . . He skirted the core areas of the adjacent packs in order to stay out of trouble with other wolves (a smart thing for a young single wolf) . . . and he moved completely around Lake Phelps before stopping on Pocosin Lakes National Wildlife Refuge." There he found a female wolf who had just paired with a sterilized, radio-collared wolf–coyote hybrid. The hybrid was soon displaced from the area and subsequently was found dead (located by his telemetry signal). Examination of his body showed that 11301M had almost certainly killed his rival. The victorious male then paired with the female, and together they will form the new Pocosin Lakes pack.

A Successful Program

In 2007, there were about a hundred red wolves, in some twenty packs, well established in the wild. Since the first were released some twenty years ago, about five hundred pups have been born in the wild population. The first experimental population release area was expanded to include three national wildlife refuges, a Department of Defense bombing range, state-owned lands, and private property—about 1.7 million acres in five counties in North Carolina, and there are red wolf release sites across 15,445 acres of private land.

In fact, the Red Wolf Field Team achieved in five years (1999 to 2004) a level of success that some scientists had believed would take fifteen. Barry Braden, who headed the US Wolf Conservation Center for three years, told me that the management teams working to return the red wolf to North Carolina as well as the Rocky Mountain gray wolf to the Northern Rockies have been successful because there has been such excellent cooperation among government personnel, nongovernmental organizations (NGOs), and concerned citizen grassroots movements. "Of course," said Barry with a laugh, "these factions do not always agree, but they all care, and they sort it out." This is in sharp contrast, he told me, with the style of the management team working with the Mexican gray wolf project—which has not so far been a success.

The team leader for the Red Wolf Recovery Program, Bud Fazio, told me of his enormous regard for the biologists who are part of his field team—some of whom, like Chris Lucash and Michael Morse, have up to twenty-one years of experience working with wolves. All are dedicated field biologists who, for nearly seven days per week and sometimes twenty-four hours per day, handle and monitor the wild red wolf population, manage coyotes, take part in education programs, talk to landowners, and resolve the many problems that crop up in a field program of this scope and complexity. The work can be physically demanding. Chris gave me an example.

"Whelping season is a brief period each spring that the field biologists both look forward to and dread," Chris said. First they must find the den following signals from the mother's radio collar (or the father's if she has lost hers). Having located the pups, they check their health, weigh them, take a drop of blood from each for a genetic record, and insert a tiny transponder chip under each pup's skin for lifelong instant identification (as we do with our dogs). It doesn't sound too difficult, but according to Chris—and this is the dreaded part—the wolves choose isolated places, as unapproachable as possible, for whelping. And "the whelping season also coincides with other uninviting seasonal changes: the beginning of high heat and humidity, the prolific growth of thorny vines and poison ivy, and the burgeoning population of biting insects."

And so, Chris continued, "For long stretches, I have to drag myself on my elbows through low narrow tunnels, through dense shrubs and downed trees overgrown with blackberry and entwined with honey-suckle, greenbrier, and grapevine, driven on by the fleeting hope of finding a den or a pup—but also by the unnerving thought of countless seed ticks traveling up my clothing and the maddening realization that dozens have already made it through to my skin."

Usually such a search takes hours, and often it is unsuccessful. The mother whom they are tracking may not be at the den; if she hears them approaching, she may lead them in the wrong direction. "Some years," said Chris, "I find nothing but lonely, empty daybeds, followed then by several weeks of itching."

Coyotes, Farmers, and Other Challenges

One major problem for the recovery plan is the migration into red wolf release sites of coyotes (not native to this part of North Carolina). This has led to two problems. There is a lot of hunting in the area, and unfortunately the coyote is becoming increasingly popular with hunters. Red wolves are sometimes mistaken for these eastern coyotes, particularly young wolves who, as was mentioned, look very similar in size and color, and this has led to a number of red wolves being shot by mistake. Thus, educating the public about red wolves is a major challenge. The second coyote-related problem is that red wolves will mate with coyotes when they cannot find a red wolf to mate with, thus creating a hybrid animal. The Red Wolf Recovery Program's coyote control strategy is attempting, with some success, to establish a coyote-free zone in and around the area into which the red wolf has been introduced.

For the most part, people have been tolerant of the return of the red wolf to its ancestral range, and fortunately the wolves are typically shy animals, and usually avoid humans and human activities. There are, of course, farmers who believe that the wolves are a threat to their livestock, but these fears have proved ill founded. During the first twenty years of the program, the wolves were seldom found guilty of killing domestic animals. There were only three proven incidents—a duck, a chicken, and a dog. And on the positive side, red wolves prey on the nutria that were introduced into the area and are a nuisance to farmers. They also hunt the raccoons who take eggs and young birds, and this may have led to an increase in bird populations, including quail and turkeys. All of this has helped to give red wolves a good reputation in the local community.

One of the most important aspects in any plan to release large predators is a good education program—and it must be prepared by people who understand and are sensitive to the concerns, fears, and prejudices of people living in the area. David Denton, hunter education specialist with the North Carolina Wildlife Resources Commission, together with the red wolf staff, works hard to ensure that people in the area understand, as much as possible, red wolf behavior and how to behave if a red wolf is encountered. They also teach hunters to recognize the difference between young red wolves and coyotes.

Howling with the Wolves

For the last ten years, the Red Wolf Coalition, the only red wolf citizen support organization, has pursued its mission of educating people by spreading awareness. Extremely popular are the "Howling Safaris": People can visit the refuge to hear the magical chorus of a red wolf pack. I remember so clearly when I first heard wolves howling in Yellowstone National Park. It is utterly unforgettable.

Field biologists sometimes howl to the wolves they know so well. "You never forget," Michael Morse wrote to me, "the first time a wild wolf responds to your howls, offered into the dark night." On his first attempt he was not an accomplished howler, and he ended with a series of uncontrollable coughs—to the great amusement of the senior wolf biologists. "But they stopped laughing when the two newly released red wolf brothers returned my howl!" said Michael. "And although my vocal cords felt scorched, the swelling sensation in my chest and mind made all else insignificant."

It did not surprise me to learn that the Red Wolf Recovery Program won America's highest conservation honor in 2007, the North American Conservation Award from the Association of Zoos and Aquariums (AZA). So many people have worked with and for the program in so many different capacities, giving so much of their lives, since it was first launched. And I know that for all of them, whether they are donors, partners, volunteers, or the biologists working long hours in often demanding circumstances, the knowledge that the red wolves are once more roaming freely in the land of their ancestors will be thanks enough. The best reward they could ask for will be the haunting sound of red wolves howling under the moon.

THANE'S FIELD NOTES

Tahki or Przewalski's Horse
(Equus ferus przewalskii, Equus przewalskii, or Equus caballus przewalskii) (classification is debated)

The first time I went to Mongolia, I said to myself, "Now, *this* is a great place to be a horse." It is a place with no fences. And no phone or electric lines. A land of beautiful and strong people who are tougher than woodpecker lips. Of course, there's not much shade. If you want trees, drive north for three days to Siberia. What makes the Mongolian steppes so famously good for horses is that this is a nation of high-desert grasslands.

And upon this shadeless grassland, with that strength Mongolians are known for, the people of this country managed to save and restore the last truly wild horses of the world. Officially, the International Union for the Conservation of Nature and Natural Resources listed Mongolia's magnificent tahki—also called Przewalski's horses—as extinct in the wild in 1968. But thanks to both captive breeding in zoos and leadership by Mongolian wildlife officials, I was able to gaze upon a restored wild herd in the summer of 2007.

My adventures in Mongolia have been in the company of a most amazing PhD wildlife biologist named Munkhtsog. Today he is one of the nation's leading scientists. It is through him that I have been able to get a glimpse of the effort it has taken to save them.

When humans first walked out of Africa fifty to seventy thousand years ago to spread across Asia and Europe, they viewed the huge herds of wild horses as prey. Eventually, of course, humans domesticated horses from wild stock, selectively breeding them for everything from transportation to work to simple beauty. However, along the way, domestication and spreading human settlements led to the extinction of the world's wild herds.

Then, to everyone's great surprise, European explorers reported seeing herds of ancestral wild horses in Central Asia. One of these explorers was Colonel Nikolai Przewalski, the Russian explorer sent by the czar on a Lewis-and-Clark-style voyage of discovery to see what was worth taking in the Gobi Desert. In 1881, Przewalski was the first to describe this mule-like horse as living in small herds of five to fifteen in the Takhiin Shar Nuru Mountains near the edge of the Gobi.

Przewalski's horses may be mule-*shaped*, but they are much lovelier creatures. They have tawny hair that is thick for the harsh winters and golden reddish in the early light of dawn, which is the best time to see them. Like many herd animals, they are naturally wary, and mothers are clearly attentive to their foals. Always on the alert, the dominant stallion will move the herd as he sees fit, but all the members stay alert for predators.

By the turn of the twentieth century, there was a frenzy among European zoos to exhibit this already rare and elusive species. Naturally, travel was harsh from southwest Mongolia all the way back to the London and Rotterdam zoos, and many of the horses perished in transit. And as fate would have it, it was a good thing that Przewalski's horse was taken into captivity. By 1968, due to hunting and habitat loss (among other factors), the species was completely extinct in the wild.

At that time it was thought that the sounds of wild horse herds would never again be heard on earth. Even in the United States, our "wild" horses, such as the mustang, were once domesticated, then escaped and returned to wild status. The Przewalski's horse was never domesticated, which is why it is considered the last truly wild horse.

Fortunately, from just a stock of thirteen zoo animals, the species has made a remarkable comeback and can again be seen in Hustai National Park, where they thrive. The horses have even become a draw for foreign tourists, as well as conservationists.

Today more than fifteen hundred Przewalski's horses live in zoos and captive breeding herds from Ohio to Ukraine, and more than four hundred roam in protected parks in Mongolia and China. The challenge, of course, is that all these animals share the genes of only those thirteen "founder" horses—the last remaining genetic stock after the species was finally extirpated from the wild. As a result, even relatively large herds are more susceptible to disease than in other, more diverse species. Fortunately, the Przewalski's horse is clearly recognized as a priority to international conservation programs, and intense cooperation continues between the managers of the captive and wild herds to ensure adequate veterinary care and genetic management for the future.

Munkhtsog was part of the team of biologists who, in 1994, re-leased the captive herd to their new home at Hustai National Park in central Mongolia. Keeping the tahki safe and thriving in the park re-mains an ongoing task—especially because they are now easy prey for wolves. (Captive-bred animals are naive about natural threats, and predation is one of the leading reasons reintroduction efforts fail.) Munkhtsog explained to me that up to 31 percent of the foals born each spring fall prey to wolves. Over time, in areas so vast, conservationists will be able to work to establish a healthy predator–prey balance again. In fact, the percentage of foal loss to wolf pre-dation is steadily, if slowly, declining.

For Munkhtsog the return of the tahki (the animals' Mongolian name) clearly is not solely about science. "The tahki is a national symbol of great pride to the Mongolian people," he said to me. "We are a nation of horsemen, and now we have proven to the world how seriously we take our horses."

One morning, after a long journey in a battered truck, bouncing along rocky, dusty roads, I finally saw the elusive, almost mythical tahki in the Mongolian steppes. Munkhtsog was with me that morn-ing, standing on the crest of a hill just after dawn.

He said that we should sit still on the grass so that we seemed less a threat to the mares with foals. And sure enough, after just an hour or so of watching, the herd of forty-three horses that had been feeding at least a kilometer away began to slowly move our way—until they were passing quite close to us. What struck me most was the beauty of the mares and their apparent concern for their young. The foals appeared oblivious to any threats, but their mothers were wary about almost anything that moved. I noted that the younger the foal, the more it looked like a domestic horse—thin-bodied and long-limbed. But the adults, and particularly the stallions, grew to be thick-bodied with proportionately shorter legs.

As I marveled at the wild herd below, Munkhtsog slapped me on the back and said, "In the US, you have thoroughbreds for racing. But in Mongolia, we have true horses!"

PART 2

Saved at the Eleventh Hour

Introduction

In this section, we find a fascinating array of different species with one thing in common—they were all brought to the very brink of extinction and have all been given a second chance. Unlike the animals discussed in part 1, none of these species were ever declared "extinct in the wild"—although all certainly would have been but for those who determined it should not be so. The restoration of these species involved taking some individuals from the remaining wild populations for captive breeding—and the critics of captive breeding were often vociferous, the proponents, as always, determined.

The story of the return of the peregrine falcon, for example, represents an extraordinary effort by literally hundreds of people across the United States. The peregrine itself was never reduced to the small numbers of the other species in this part, but it was totally extirpated from a huge part of its original range in the eastern United States. And the account of the battle to ban the use of DDT is chilling in its revelation of the determination of major corporations to trample over other life-forms in their quest for wealth. The winning of that battle was a triumph for the environmental movement, and helped to save countless other species in addition to the peregrine.

These are our first stories about those dedicated to protecting not just the charismatic animals, but fishes, reptiles, and insects. "Why on earth," people ask, "would anyone devote themselves to protecting a bug? The world would be better off without them." When I was a child, we had a painting on the wall showing a cute little girl cuddled up with a bulldog of somewhat scowling aspect. It was captioned, "Everyone is loved by someone." And the people whose stories we share here do care passionately about the creatures they are trying to protect. But they know, too, that every species has its own unique niche in the ecosystem—that interconnected web of life—and as such is important. This is one reason why the costs, sometimes great, are truly worth it.

And it is important to recognize that the animal species with which we share the planet have value in their own right. We have messed things up for so many—it is up to us to put things right.

Golden Lion Tamarin
(Leontopithecus rosalia)

The first time I met a golden lion tamarin face-to-face was at the National Zoo in Washington, DC, on a beautiful spring morning in 2007. I also met Dr. Devra Kleiman, who had kindly offered to share some of her vast knowledge of the species to which she has devoted much of her life.

In the early 1800s, golden lion tamarins were apparently common in the Atlantic Coastal Forest of eastern Brazil, but their number was drastically reduced throughout the second half of the twentieth century as they were captured for exotic pets and zoos, and their forest habitat was destroyed to give way to pasture for cattle, agriculture, and plantation forestry. Today less than 7 percent of the original Atlantic Forest remains, much of it fragmented.

Rescued by Brazil's Father of Primatology

There are four species of lion tamarins: the black lion tamarin, *Leontopithecus chrysopygus;* the golden-headed lion tamarin, *L. chrysomelas;* the black-faced lion tamarin, *L. caissara;* and the golden lion tamarin, *L. rosalia.* The golden lion tamarins are among the most endangered of all New World primates. They might have vanished altogether but for the dedication, passion, and persistence of Dr. Coimbra-Filho—often called the Father of Primatology in Brazil—and his colleague Alceo Magnanini.

As early as 1962, these two scientists recognized the need for a breeding program for golden lion tamarins, with the goal of reintroducing them into protected forests. But they got little support, and the attempt to

Devra Kleiman at the small mammal house of the National Zoo, checking on this golden lion tamarin's climbing abilities before its release into the rain forest of Brazil. *(Jessie Cohan, Smithsonian National Zoo)*

start the facility failed. However, they continued their work throughout the 1960s and 1970s and, mostly using their own money, traveled to many municipalities in search of the tamarins, visiting villages and interviewing the local people, especially hunters. The work was hard and often depressing. They identified two areas that would have been ideal sites for reintroduction—but both had been destroyed, along with countless other tracts of forest, when they returned a year later.

Difficult times indeed, yet extraordinarily valuable, for they gathered data confirming the desperate plight of the lion tamarins and their habitat, which was essential for their battle to save them. And they earmarked an area of forest that, due to the persistence of Dr. Coimbra-Filho, eventually became the Poço das Antas Biological Reserve, created for the purpose of protecting the golden lion tamarins. It was the first biological reserve in Brazil.

In 1972, a groundbreaking conference titled Saving the Lion Marmoset (as they were called in those days) brought together twenty-eight biologists from Europe, America, and Brazil. It focused international attention on the urgent need to prevent the golden lion tamarin from sliding into extinction. Plans were drawn up for conservation in the wild, support was obtained for Dr. Coimbra-Filho's breeding program in Brazil, and a strategy was created for a coordinated global captive breeding program in zoos. It was this conference that led to the Golden Lion Tamarin Conservation Program at the National Zoological Park in Washington, DC. And it was at this conference that Devra began her long involvement with the little primates.

Meeting a Golden Lion Tamarin Family

My visit to the National Zoo took place thirty-five years after that conference. I had never met golden lion tamarins close up, and it was a real treat to go with Devra and their keeper, Eric Smith, into the newly constructed enclosure of a family group. There I met the adult pair, Eduardo and Laranja; two adolescent females, Samba and Gisella; and two youngsters, Mara and Mo. I was enchanted. They are like living jewels of the deep forest with shining golden hair that cloaks their bodies and frames the face with a leonine mane. As I watched them, slightly apprehensive with so many strangers in their new home, I felt a surge of gratitude for all the hard work and tears that had prevented their extinction.

Afterward a small group of us gathered to talk tamarin. I asked Devra how she got involved. She told us she grew up in suburbs of New York, with no nature and no pets, destined for medical school. Then as part of a college project, she observed a wolf pack in a zoo, became fascinated, and realized that she wanted to study animal behavior. Interestingly, she spent time at the London Zoo and worked with Desmond Morris— just as I had. She specialized in the comparative and social reproductive behavior of mammals and worked with many species—until she learned about the plight of the golden lion tamarin.

"I was determined to do my best for those enchanting little creatures," she told us. So she set to work to raise money, gather information, and start a coordinated breeding program. Many people believed such a scheme would never work. Smiling, she recalled the advice given her back then: "Don't get involved with tamarins. They are going extinct—it will be bad for your career.

"I am so glad I didn't follow that advice," she added. Indeed, it was fortunate for all of us, especially the golden lion tamarins!

Devra contacted all the zoos that kept golden lion tamarins and found that almost nothing was known about tamarin reproductive behavior. "No one even knew whether they should be kept in monogamous or polygamous breeding groups," she said. But eventually she came to believe that tamarin groups in the wild, containing two to eight individuals, might be composed of a mated pair and their offspring. So she recommended that the zoo keep adult pairs on their own, so that family groups could form naturally. This was the key to success. Over time, as more became known about the tamarins' natural diets and social system and was applied to their care, the situation improved. But even so, by the end of 1975, there were still only eighty-three golden lion tamarins spread through sixteen institutions outside Brazil and another thirty-nine individuals at the facility in Brazil.

Return to the Wild

Gradually, though, the captive population grew, and Devra began to concentrate on the next stage—returning the species to the wild. The first step, of course, was to find a safe environment for them. "I traveled to Brazil to visit the reserve where it was hoped the tamarins would be released," Devra recalled. "The Atlantic Coastal Forest there had been decimated, and even when we got to the reserve, there was still very little forest remaining. To my horror, the guard at the gate to the reserve had a pet tamarin on a leash! It seemed impossible that we could do a successful reintroduction there. But that was all that was left of their natural habitat. We would have to work with what was there."

The scientist and conservationist Dr. Benjamin Beck was selected to take charge of coordinating the release program. First the groundwork had to be laid. Devra and Ben made repeated trips to Brazil, developing close relationships with their Brazilian colleagues. By 1984, all was ready: A release area had been secured, Brazilian partners and staff acquired. The first captive golden lion tamarins were released in the forest.

"We realized after that first release," Devra told us, "that the captive-born animals had problems moving about in the trees; they simply did not know how to navigate complex 3-D environments." But they managed somehow, and at the same time the team was learning a lot about their behavior. One day, Devra told me, she was following an adolescent female and her two young brothers, Ron and Mark, who had separated from the rest of the group. Farther and farther they wandered, exploring their new world, and as dusk fell Devra feared they might be lost. But suddenly the female gave a strange call and headed off with great purpose, calling as she went. Ron and Mark immediately followed—and Devra followed them. "I almost felt like part of the family," said Devra. "We all kept up, following the calls." And in less than thirty minutes, they were back at the nest box. Subsequently, the researchers learned that this call means "Let's go!" They named it the "vamonos call."

Adapting to the Forest

Soon after this, Devra and Ben made a bold and innovative decision— they would allow some of the golden lion tamarin families to roam freely in a small patch of forest on the grounds of the National Zoo in DC. This would allow them to become familiar with treetop travel before being released in Brazil. The plan, under Ben's direction, was a success. "For one thing," said Devra, "once they were outside, they instinctively began giving the soft 'vamonos' calls that I had heard in the wild. It was wonderful!"

Not only did the tamarins learn climbing skills, but family groups established small territories—about a hundred square yards—just as they would in the wild. Devra and Ben felt, therefore, that it was unlikely any of them would leave the zoo grounds. To their great relief, they were proven right.

Ben told me that what interests him most about the release program in Brazil is that pre-release training (such as learning to poke food from crevices with their fingers or how to open fruits) does not make much difference to the golden lion tamarins' survival in the wild. What is important is the soft-release method. This means that they are provided with food and shelter when they start their life in the forest, but as they begin to eat natural foods, field researchers progressively feed and observe them less: from daily visits they cut down to three days per week, to once a week, then once a month. If an individual is hurt or gets lost, it is captured and treated before being returned. All groups have become independent after five years. The key for success, Ben explained, is for the females to live long enough to reproduce. Young tamarins, born in the wild, will do fine. "Because then," said Ben, "they are born with wild brains."

More Stories from the Wild

I asked Ben for an anecdote he could share. He told me about Emily, who arrived with four of her family in 1988. They were taken into the forest and introduced to their nest box fixed up in a tree. On the second night, it was very cold and wet. Emily seemed confused. She climbed to the very end of a branch and there she sat, huddled in the rain. Ben and his colleague Andreia Martins also sat, huddled, watching her. Eventually, it began to get dark, and in the end they were forced to leave her, small and bedraggled on the end of her branch, with the rest of her family all cozy in the nest box.

It was a subdued group of humans that gathered for supper, cold and wet themselves. "None of us slept very much," said Ben. They went out very early the next morning. When they reached the tree, Emily was lying on the ground but still alive, though extremely cold. Andreia put Emily under her shirt and took her back to camp. Gradually, Emily warmed up, and by day's end she was dry and fluffed out. She not only survived but went on to have several babies. "She was a real sweetheart," said Ben.

One day, Emily and her son disappeared. Unfortunately, some people steal the tamarins to sell them (illegally) as pets—over the years, at least twenty-two have been stolen. Amazingly, they got Emily back when a veterinarian noticed her tattoo and realized she had been stolen.

Emily soon settled down and had another family. Almost unbelievably, she was stolen again, and again they were able to get her back!

A Name or a Number?

Ben told me that they no longer give the tamarins names in the field, just numbers. This business of identifying individuals by name or number has had an interesting history in the tamarin project. "I began by giving the tamarins numbers, which seemed more scientific at the time," Devra recalled, "but to spite me, David Kessler [one of her colleagues] named a hand-reared tamarin Colonel Ezekiel Atlas Drummond—and it stuck. We have been using names ever since."

Although the captive breeding program still uses names, they switched to numbers in the field. Not because it is more scientific, but because such a relatively large percentage of the tamarins don't make it—about 80 percent are dead or have disappeared by the end of the second year in the wild. Those working with them find it less distressing if they are not known by name.

When the team finds an unmarked tamarin out in the forest, they know it marks a success story—an individual born in the wild who has sought out and established its own territory. Some have even made it across more than a mile of open agricultural land. The team no longer spends time observing the family units closely. Occasional monitoring of their health, reproduction, and survival rate is all that's required.

Meanwhile, as the introduced tamarins thrived, there were still some highly endangered groups of wild golden lion tamarins. An exhaustive survey in the early 1990s had revealed that there were sixty individuals in twelve groups, living in nine very small fragmented patches of forest that were destined to be cut down to build beach condominiums. And so, between 1994 and 1997, six of the groups (forty-three individuals) were translocated to what is now the União Biological Reserve.

Key to Long-Term Success: Handing Over to the Brazilians

From the beginning, Devra knew that a key component for the success of the golden lion tamarin reintroduction program would be the attitude of local farmers—those with remnant forest into which the growing numbers of family groups could be reintroduced. And so from the earliest days, the Brazilian team worked on forging relationships with the local people. It was hard going at first, for many of the farmers were initially hostile, Devra told us. "But it was perhaps the most important aspect. I wanted to be able to retire and know that there was something in place that was lasting, and this could only be possible if it was in Brazilian hands."

To a very large extent, this has now happened. In 1992, the Golden Lion Tamarin Association (or the Associação Mico-Leão Dourado—AMLD) was formed in Brazil to integrate all conservation work relating to the golden lion tamarins and to educate local communities about the conservation program. The association, headed by a dynamic young Brazilian, Denise Rambaldi, monitors the tamarin populations, helps impoverished farmers develop agro-forestry techniques, and trains young Brazilians in conservation. The association also works closely with Brazilian government agencies to foster conservation in the entire region.

In 2003, the golden lion tamarin was downlisted from critically endangered to endangered on the IUCN Red List of Threatened Species, the only primate species to have been downlisted as a result of a conservation effort. This is certainly a milestone for the countless people and organizations who have dedicated themselves to the species' survival.

Of course, as with all conservation projects, those who care cannot sit back and relax. Habitat is still being destroyed, and the continuing fragmentation of existing forests remains the tamarins' greatest threat to survival. Thus it is very encouraging to learn that the AMLD is building forest corridors to link tamarin habitats, which will help prevent inbreeding within small isolated groups. The first of those corridors, which will be approximately twelve miles in length, is nearly complete. And more and more private ranchers are agreeing to accept tamarin groups on their land.

At the time of writing, there are golden lion tamarins living on twenty-one private ranches adjacent to the Poço das Antas Biological Reserve. When their currency was redesigned, the Brazilian people voted to portray the golden lion tamarin on twenty-dollar banknotes—the species is now an icon of conservation in Brazil.

"When I started working with the zoo population in 1972, there were about seventy golden lion tamarins in zoos," Devra said. By the late 1980s, that number had increased to almost five hundred, and it was decided to put some individuals on contraceptives and stabilize the captive population. Today there are about 470 in zoos and aquariums, and the groups are carefully managed. "In 1984, when I started reintroducing tamarins, there were fewer than five hundred in the wild," Devra told me. Thanks to the reintroduction efforts, about sixteen to seventeen hundred tamarins now live in the wild.

As I write this, in my home in faraway Bournemouth, I think back to that April day when Devra introduced me to Eduardo and Laranja and their family. I remember how the adult male approached Devra, who had been handed a piece of banana by the keeper. Gently, the small creature reached out to take the fruit. It was, for me, a magical moment, symbolizing the trust of a very small primate for the woman who has worked so passionately to prevent this enchanting species from vanishing forever from Planet Earth.

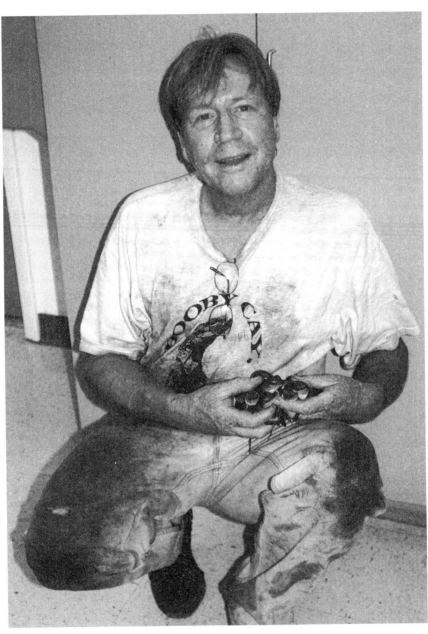

Joe Wasilewski, who is helping to ensure the future of the American crocodile, with three wild hatchlings at Turkey Point Nuclear Power Plant, 2007. (Joseph A. Wasilewski)

American Crocodile
(*Crocodylus acutus*)

For most people—including me—the thought of encountering a crocodile in the water is quite terrifying. I vividly remember empathizing with the elephant when my mother read me that most delightful of Rudyard Kipling's *Just So Stories:* "How the Elephant Got His Trunk." The poor little elephant child wanders down to the "great grey-green greasy Limpopo River" to drink, only to have his short little nose seized by a crocodile. The crocodile pulls and pulls and the elephant pulls and pulls. Luckily all his uncles and aunts hurry to the rescue. They pull and pull, and the crocodile pulls and pulls, until by the time he is rescued, the nose of the elephant child has been elongated into a trunk.

In real life, there are fearsome accounts of large antelopes—even buffalo—being seized by crocodiles as they go to drink; struggling desperately, they are pulled under the water to their deaths. When first we arrived at Gombe, my mother and I were warned about two such crocodiles that frequented the lakeshore near our camp. Nothing would have induced either of us to swim in the lake in those days. Indeed, one of those crocs almost grabbed the cook's wife. Later we were told they were the "familiars" (like the black cat of a witch) of old Iddi Matata who, although we had no idea at the time, was the most infamous witch doctor in the area. And it is true that when he moved away, those two crocodiles disappeared. Indeed, there are many stories about crocodiles in association with the powerful witch doctors of Tanzania.

A "Gentle" and "Timid" Crocodile

But all of those stories are about *African* or *Nile* crocodiles, who behave much like the American alligator. In this chapter, we shall hear about the *American* crocodile. This is a very different kind of animal—much gentler and more timid, but unfortunately often feared and persecuted by those who mistake it for an alligator. Once you know the difference, though, it is easy to distinguish between the two. First, the crocodile is olive green to gray-brown, mottled with black, whereas the alligator is uniformly black. Second, the crocodile has a much narrower snout, and the fourth tooth on the bottom on each side of its mouth is clearly visible on the outside of the upper jaw. There has never been a documented attack on humans by the crocodile in Florida, although we hear that there have been a few in Mexico and Costa Rica.

The American crocodile has a large range, including Cuba, Jamaica, Hispaniola, the Caribbean coast from Venezuela to the Yucatán, and the Pacific coast from Peru to Mexico. The northern subspecies found in Florida has been isolated from its relatives for at least sixty thousand years (although recent but unpublished DNA studies show relatively recent mixing with the American crocodiles of Cuba). By the early 1970s, the Florida subspecies, like many other crocodilians around the world, had been driven toward extinction through hunting for its hide and relentless human development that had destroyed huge areas of wild habitat. In 1975, it was classified as endangered: It was estimated that no more than two hundred to four hundred individuals survived.

In November 2006, I had a wonderful telephone conversation with Frank Mazzotti, a University of Florida wildlife biologist who has been involved with crocodile research for almost thirty years. In 1977 Frank, then a graduate student, began assisting with fieldwork on the crocodile in Everglades National Park. No one knew much about it except that it appeared to be in dire straits. One of the questions the researchers were trying to answer was: How many of the young crocs were surviving, and what was killing them?

Frank occasionally saw blue crabs eating a young croc, but thought they had probably scavenged a dead individual. Then one memorable day he saw a reptilian tail thrashing in the water, grabbed it, and pulled out a young crocodile that was firmly in the grips of a blue crab, which had one claw around the middle of its prey and the other around its head. Frank managed to free the youngster, but it was no longer breathing. A short while previously, someone had pinned a silly cartoon from *MAD* magazine to the wall in the ranger station.

"It showed a guy giving a lizard mouth-to-mouth resuscitation," said Frank. "The character just closed his lips around the lizard's neck and blew." So that is what he did to the crocodile! And after a few seconds it spat up water and, thoroughly revived, was soon ready to go. Surely, Frank is the only person in the world who has given the kiss of life to a crocodile!

A Love of the Wilderness at Night

When he was a child, Frank read all the Tarzan books and other similar stories—just like me. But whereas I fell in love with Tarzan, he wanted to *be* Tarzan. "Gradually," he said, "I realized that this would not happen to a five-foot-eight-inch adolescent!" But as a college student, he had the opportunity to help with some crocodile research. Since the animals are nocturnal, it meant being in the wilderness after dark—which he loved. At that time, they kept a few young crocs as part of the research program. "I did raise a couple until they were about six feet long, and got to know them pretty well until they had to be released." When he talks about them, he cannot keep the enthusiasm out of his voice. "They are the real sweethearts of the crocodilian world. The least defensive, and so the least aggressive. They are shy," he said. "And they are relatively gentle." At this point he laughed—we decided the crocodile might not seem so gentle from the vantage point of its prey! (It feeds on crabs, fish, snakes, turtles, birds, and small mammals, rarely anything larger than a raccoon or rabbit.)

After completing his PhD, Frank conducted a survey of crocodile nests in Florida. First he pulled together all the information he could find, starting from 1930, as to where nests had been recorded. Then he visited each of the sites, and searched—in vain—for signs of crocodiles. Finally, in 1987, he found a nest at Club Key in Florida Bay, Everglades National Park. "The last one recorded there had been in 1953," he told me. It was nearly twenty years since he discovered that nest, but the excitement in his voice came down the phone line all the way from Florida to my home in Bournemouth!

Crocodile Mothering

I was fascinated to learn that female crocodiles (just like chimpanzees!) do not become sexually mature until they are eleven to thirteen years of age and (also like chimps) can live for about sixty years. After mating, which occurs in late winter and early spring, the female crocodile digs a nest hole on high ground such as a beach or streambank, lays twenty to fifty eggs, and carefully covers them with soil. She then leaves the nest, but after about eighty-five days she returns—for this is when the babies are due to hatch, and they will need her help to dig their way out. When she arrives at her nest, she puts her ear to the ground to listen for the chattering sounds the hatchlings make when breaking out of their eggs. Then she uncovers them and carries them to the water in her mouth. On their own, the nine-inch-long youngsters make their way into saltwater estuaries. Frank told me his favorite memories are of this maternal behavior.

The survival rate of the young crocodiles for the first year ranges from 6 to 50 percent and depends, in part, on the amount of rainfall and natural water flow—they cannot tolerate high salinity. Historically, fresh water flowing through the Everglades lowered the water's salinity where it emptied into Florida Bay, producing the conditions young crocodiles need. The problem, of course, is that the natural water flow was disrupted long ago. For the past few decades, water has been "managed"—held in catchment areas outside the park for agricultural uses then, when it is no longer needed, suddenly released in large quantities. This has disrupted the slow, relatively constant flow of fresh water through the glades, affecting water levels in the wetlands and the salinity of Florida Bay, wreaking havoc on both flora and fauna.

How a Power Station Helped Save the American Crocodile

Despite this, scientists estimate that there are about four times as many crocodiles in Florida today as there were in 1975. The extraordinary thing is that this population growth is in large part due to the operations of a power station! In the 1970s, Florida Power and Light, at Turkey Point, built 168 miles of canals that allow the water coming from the plant to cool before reentering the bay. And this created ideal habitat for crocodiles, which dig their nests in the loose soil between the canals. To its credit, when hatchlings were discovered there in 1978, the company showed great interest and hired a consulting firm to monitor the animals. Since then, the number of nesting crocodiles has steadily increased.

For information about the crocodile situation at Turkey Point, I turned to Joe Wasilewski, who started working there in 1996 and has been there ever since. Part of his job is to hunt for the crocodiles, following every print and "tail-drag" he sees in the warm, saline-rich canals. Every crocodile he catches is tagged with a microchip.

"I've caught thousands of them," Joe said during a phone call in spring 2008. "They're not that aggressive. Once they know they've been had, they surrender. Whereas other species of crocs will fight till the bitter end."

The population of American crocodiles—excluding hatchlings—within and adjacent to the FPL Turkey Point cooling canal system has been surveyed, using the same methods, since 1985. The results show a dramatic increase. That first year there were just nineteen individuals, ten years later the number was forty, and by 2005 the population had increased to four hundred.

The crocodile also nests in the southern mainland area of Everglades National Park and in the sixty-six-hundred-acre Crocodile Lake National Wildlife Refuge in Key Largo established in 1980. Thus 90 percent of its habitat is either in protected areas or on land owned by a very supportive company. And with added protection and growing numbers, the crocodiles have been showing up in populated areas—from inland waterways to golf course ponds. For this reason, as Frank said, it is very important to educate people about the American crocodile's passive nature, and teach people how to distinguish it from the much more aggressive alligator.

The Usefulness of Crocodiles

The crocodiles play an important and interesting role in the ecosystem. "For instance," said Joe, "we've had a horrible problem with invasive species here in Florida—exotic pets, such as green iguanas and pythons, get released into the wild. Fortunately, the crocodiles are an apex species—they eat anything smaller than themselves and so help to keep the invasive species under control!"

Ironically one of the signs of a healthy crocodile population is that they start preying on their own young. "As the numbers grow, they take on their own population control," Joe said. "We're seeing more and more crocodiles eating the hatchlings. Some just get a taste for it."

So far, the comeback of the American crocodile can be considered a success story. Its ultimate fate, however, depends—like that of so many other Florida flora and fauna—on the restoration of the Everglades. We have to hope that engineers, working with biologists, will succeed in ensuring a more natural water flow. The passion and persistence of Frank Mazzotti and Joe Wasilewski—and the presence of the crocodile itself—may make all the difference.

Falconer Tom Cade, who led the massive American efforts to restore the peregrine falcon to its original hunting grounds. *(J. Sherwood Chalmers/The Peregrine Fund)*

Peregrine Falcon
(Falco peregrinus)

The first time I watched a peregrine falcon streak across the sky, then dive down after some small bird of prey, I experienced a similar tingling of wonder and magic as when I see a shooting star. All bird flight is awesome—no wonder we earthbound humans strove for so long to find ways to fly. No wonder that most of us, at one time or another, dream that we are flying. (Indeed my mother's dream was once so vivid that she rose and, still half asleep, launched herself from the end of her bed—waking the whole household with her inevitable heavy landing!)

As a child, I read a story about a little boy and a falcon—I don't know what species. They loved each other so much that he never had to hood or shackle her. She hunted to get food for both of them when, for some long-forgotten reason, he was hiding in the moors. I have always been ambivalent about falconry—the taming and restriction of birds, symbols of freedom shackled. That is why caged birds, denied their birthright, make me sad, angry. But I am filled with admiration for the role falconers played in the restoration of the glorious peregrine falcon as described in this chapter. Indeed, Tom Cade, the man who initiated and led the effort, is a falconer himself, and he relied heavily on the knowledge and skills of his fellow enthusiasts.

The peregrine (along with the gyrfalcon) has long been considered the classic bird of the ancient art of falconry. The American naturalist Roger Tory Peterson wrote, "Man emerged from the shadows of antiquity with a Peregrine on his wrist." Although peregrines were sometimes used simply to hunt for the pot, for the most part falconry has been a sport for nobility. It did not start in North America until the 1900s, and the peregrine quickly became the favorite bird there also.

Much has been written about this falcon—its beauty, its speed, its deadly stoop as it dives upon its prey from above. However, one book, *Return of the Peregrine: A North American Saga of Tenacity and Teamwork*, documents its near extinction in America and the incredible story of its rescue and return to the wild. This book, along with personal information from Tom Cade, who led the restoration effort, has been our main source of information for this chapter.

Peregrines have been observed in almost all parts of the globe save Antarctica. They were always more abundant in Europe than in North America—indeed, during World War II, many peregrine falcons in the south of England were killed for the risk they posed to service carrier pigeons. But after the war, some ornithologists began to suspect that all was not well with Britain's peregrines. And in 1960, the Bird Trust for Ornithology (BTO) asked the late Derek Ratcliffe (the chief scientist for the British government's Nature Conservancy Council) to make a survey of nesting peregrine falcons throughout the UK. He found that numbers were indeed in serious decline in the south, reduced in the rest of England and Wales, and only normal in remote areas of Scotland.

The BTO suggested this was perhaps due to the very toxic organo-chlorine pesticides that had been introduced to British agriculture after World War II. There had been many reports, dating from the late 1940s, of seed-eating birds dying seemingly as a result of feeding on pesticide-treated fields. Corpses of raptors were also discovered, presumed poisoned from feeding on contaminated prey.

Ratcliffe's next survey, in 1963, showed a further dramatic drop in peregrine numbers, especially in the south, where only three pairs were found. Again, it was only in remote areas of Scotland that the birds were unaffected. Reports coming in from Europe documented a similar decline in bird populations, including peregrines.

The Fight to Ban DDT

As more people became aware of the death of so many birds, there was a great public outcry. The British government tasked scientists at Monks Wood research station with conducting studies on the adverse effects of pesticides. Meanwhile a series of "voluntary bans" were recommended, restricting the use of organochlorines and other toxic pesticides. Ratcliffe had found that even occupied eyries (nests of eagles and falcons) often contained broken eggs. Suspecting that chemicals were affecting the thickness of the eggshell, he took an addled egg to Monks Wood for testing. It contained traces of DDE (the residual product of DDT) and other chemical pesticides.

Thus the British team had already been studying the effects of chemical pesticides when Rachel Carson published *Silent Spring* in 1962—the results of her own research into why thousands of birds and insects in the United States were also dying. And reports coming in from Europe documented a similar decline in bird populations, including peregrines, also thought to be caused by pesticides.

In the United States, Joe Hickey, a professor at the University of Wisconsin, was one of a number of ornithologists and falconers concerned by a seeming decline in the peregrine populations. In 1939, he had made an extensive survey of active nest sites of the Appalachian peregrine east of the Mississippi. In 1963, he enlisted Daniel Berger (who had been making annual surveys of the peregrine population along the Mississippi for thirteen years) to undertake a survey of nest sites in the areas he (Hickey) had covered more than twenty years before. In 1964, Berger and his teammate traveled to fourteen US states and one Canadian province. They found no occupied eyries and saw not one peregrine for the whole three months. The crash of the Appalachian peregrine population in the eastern US was total.

Soon after learning this shocking news, Hickey heard about the pesticide situation in the UK from Ratcliffe. He at once set about organizing a gathering to which all interested parties were invited: falconers, scientists, government officials, even representatives of the agriculture and pharmaceutical companies. The Madison Conference, as it came to be known, took place in Wisconsin in mid-1965. There Ratcliffe explained to the assembled group what was going on in the UK. And he reported on a conference he had just attended involving seventy-one scientists from eleven countries in Europe, who had concluded that persistent pesticides in general, and the organochlorines in particular, posed a major threat to wildlife.

"Almost immediately after the conference people started looking at eggs and also tissues of Peregrines that had died—and they found both DDT and the residual product DDE," Tom Cade wrote. "From that point on, it was pretty clear that DDT was the main problem the Peregrines were facing." But more scientific "proof" was needed to convince

governments to legislate against the use of these poisons given the determined opposition from the agrochemical and agricultural industries.
These interests claimed that causal correlation between thinning of egg-
shells and the use of some pesticides was circumstantial.

So Ratcliffe designed a way of calculating the thickness of an eggshell
based on its weight, length, and breadth as measured with calipers and
used it to examine eggs in collections throughout the UK. He found a
marked decrease in thickness from 1947 onward. Meanwhile, at the
Patuxent Wildlife Research Center of the US Fish and Wildlife Service,
scientists were investigating the effects of DDT and DDE on different
bird species, including kestrels. Their experiments showed that relatively small amounts of various chemicals could lead to eggshell thinning in kestrels, and that this correlated with the situation of wild
kestrels contaminated with pesticide-affected prey.

"Restrictions on DDT Will Never Happen"

The scientific evidence was mounting, but opposition remained strong.
Indeed, one of President Lyndon Johnson's scientific advisers at the
Madison conference stated that "restrictions on the use of DDT will
never happen."

"That statement," Tom told me, "served as a challenge for many
of us."

Pesticide regulation was the responsibility of the newly formed Environmental Protection Agency (EPA), which, under a court order in-
stigated by the Environmental Defense Fund, started hearings on
DDT in mid-1971. They called 125 witnesses over eight months, after
which William Ruckelshaus, appointed first EPA administrator by
President Nixon, courageously banned DDT nationally, with the president's support.

The inevitable appeal was overturned by the US Supreme Court. The
evidence given by so many scientists on the effects of the chemical on
birds of prey, particularly the peregrine, bald eagle, osprey, and brown
pelican, could not be refuted. It was a long, hard fight, but it resulted in a
major victory for conservation, and set legal precedents in environmental
law that have had far-reaching benefits for the environment.

In Canada, DDT had already been banned, and in several European
countries conservationists had also been lobbying for legislation against
the use of DDT and other harmful chemicals. In Britain, the govern-

ment's voluntary bans on the most toxic chemicals used in pesticides, along with farmers' reductions in the amounts they used, meant that most had already been phased out by 1979, when the European Union finally banned their use.

Discovering the Nature of Peregrine Breeding

In anticipation of an eventual ban on the use of DDT, Tom Cade founded the Peregrine Fund in 1970 and began plans for a captive breeding program with the goal of reintroducing peregrines into the eastern United States. However, very little was known about breeding them in captivity, although a handful of falconers in the US and Europe had achieved some success. In the late 1950s, when some falconers had already noticed that peregrine populations seemed to be in trouble, they formed the North American Falconers Association, and forty-five falconers from many states had attended the founding meeting in 1961 to discuss the situation. Some had suggested captive breeding.

Tom is a falconer himself, and again and again as he moved ahead with his ambitious plan, he sought advice and help from other falconers. It was thanks to them that he knew that young peregrines can learn to hunt without parental tutelage. And falconers had practiced "hacking" since the Elizabethan era: They would haul a hack—a kind of wagon—to the top of a hill and put falcon chicks on it just before they were able to fly. Food was delivered each day, and after fledging, the chicks could come and go at will. When they had developed muscle tone and could catch birds for themselves, they were recaptured for training. Clearly, hacking would be part of Tom's reintroduction plan. Most importantly, falconers knew that the nature of the peregrine was well suited for breeding in captivity. Tom wrote, "Although master of the sky and a denizen of wild and haunting landscapes, the Peregrine has also for centuries been . . . a bird that because of its gentle and placid disposition comes readily to the hand to do man's bidding. . . ."

Because he taught at Cornell University and was associated with its famous Laboratory of Ornithology, Tom was able to establish a breeding facility there, affectionately known as Peregrine Palace. For some of his first birds, he turned to Dr. Heinz Meng, a professor at the State University of New York–New Paltz. A keen falconer and a Cornell alumnus, Meng had set up his own small breeding program, and he lent his breeding pair, and their offspring, to Tom's program.

Through trial and error, Tom and the Peregrine Fund staff learned how to breed the greatest number of birds in the shortest time. The program incorporated natural breeding by pairs of falcons, double- or even triple-clutching, and artificial insemination, with the resulting eggs hatched either by nesting peregrines or in incubators. Gradually they learned that success depended primarily on the age at which a bird entered captivity (older individuals seldom bred) and how the nestlings were handled. They found that it was best to raise chicks in groups and take them, at five weeks of age, to the hacking boxes that were being set up in suitable places throughout the country.

Phyllis Dague and Jim Weaver, in particular, played key roles throughout the program. "Those two ran the program at Cornell," Tom told me. "Phyllis did everything—secretary, accountant, fund-raiser, baby bird feeder, field assistant." For several years, Phyllis actually lived at Peregrine Palace since Tom felt that someone should be with the birds at all times. Initially there was not even a window in the place, and in *Return of the Peregrine*, Phyllis describes dark windy nights spent on her own, living in the Peregrine Fund's "office." Indeed, that office, despite its annual condemnation by fire marshals, was used by a small group of people to accomplish great things.

Jim Weaver was also recruited in the early days at Cornell. Tom told me that Jim had a wonderful talent for handling the birds and keeping them fit in captivity. Even more important, he was a great manager and great team leader, and he recruited a team of loyal and dedicated co-workers. One was Bill Burnham who, after years of working on the restoration program, eventually founded the World Center for Birds of Prey. He was president of the Peregrine Fund until his premature death in 2006 at the age of fifty-nine.

Peregrines in Love

One of the things I have enjoyed when researching this story is the descriptions of the different peregrine personalities. I loved reading about a particularly feisty male in the Cornell breeding program named Sergeant Pepper. He terrorized rather than bonded with the females offered to him as mates. But then, after rejecting eight females in succession, "he fell in love with a little Latin lady from Chile," wrote Tom. The two birds immediately accepted each other. "They started courting and he began feeding her. And even though she came to us in the middle of her molt, she somehow accelerated her molt and came back into breeding condition that spring. And they produced a lot of young every year after that."

Artificial insemination, or AI, is considered a necessary tool only when a female refuses to accept a male's courtship—or when a male refuses to mate with *any* female. Such a male was BC (Beer Can), who was collected from the wild when he was two days old. BC was hand-reared and imprinted on humans, and he utterly rejected females. When he became part of the breeding program, therefore, he had to become a semen provider for AI, and William Heinrich was given the job of "stripping" BC by hand. This was stressful for BC—"and very undignified!" said Heinrich. So when he heard that Les Boyd had designed a "copulation hat," he persuaded him to come and explain how it worked.

Les told Heinrich to climb up to one of the nesting ledges in BC's chamber, carrying a dead bird. When BC flew to take the offering, Heinrich had to make eye contact while imitating the *ee-chip* courtship call, then bow so that his head was level with the ledge—thus enabling BC

Propagation biologist Cal Sandfort patiently wearing a "copulation hat." According to Bill Heinrich, Cal has probably raised more captive-bred peregrines than anyone in the world. *(Peregrine Fund File Photo)*

to copulate with his hat. Heinrich accordingly *ee-chip*ed and bowed: BC merely fed contentedly on the dead bird. Les instructed Heinrich to repeat the whole performance—from making eye contact to bowing—until BC showed some interest. When Heinrich was patiently doing this for the tenth time, Les could not stop himself from bursting out laughing—and Heinrich, thinking he had been made to look an idiot, climbed down and said he was quitting.

It took some time before a contrite Les was able to convince him that he had been really close to success and that his antics—while hilarious for a human observer—were far from absurd so far as BC was concerned! And so Heinrich carried on, repeating his "hilarious" courtship three times a day. And two days later, BC became the first voluntary semen donor to the Peregrine Fund. From then on, he provided good-quality semen several times a day, willingly—and perhaps joyfully! The results traveled far and wide across North America in the form of the scores of young birds he had, unknowingly, sired.

Return to the Skies

In 1974, the Peregrine Fund sent the first four youngsters from the breeding program for experimental release into the wild. Two were fostered to a wild pair in Colorado that had lost one lot of eggs (due to thin shells) and was incubating a second, dummy clutch (provided by the Peregrine Fund). These dummies were exchanged for the two captive-born chicks—they were accepted and raised successfully. The other two captive-bred chicks went to Heinz Meng, who had built a hacking facility on top of a ten-story tower on his university campus; they, too, fledged successfully. These were the first experimental releases of captive-born peregrine falcons in the United States.

The following year, sixteen chicks were sent to five hack sites in different places. Many of these young birds returned to the hack sites, or nearby, the following year. "When so many individuals came back in 1976," said Tom, "I was fairly confident that we could successfully release more, and that we had a good chance of bringing this species back."

During that period, Tom told me, "falconers from around the US lent me their peregrines to contribute to the breeding effort." They also shared hacking techniques long known to falconry, which proved invaluable for the reintroduction to the wild. Indeed, all releases were initially carried out by falconers, although, as the project became widely known, hundreds of volunteers from all walks of life supported them as "hack site attendants." This was a tough assignment, involving camping out at hack sites for weeks, enduring heat, cold, and insects, not to mention close encounters with bears and moose, bites from rattlesnakes, and even wildfires. Yet almost all accepted the hardship uncomplainingly, and developed a deep respect for the birds whose future they were helping to assure. "A Peregrine owns the air like nothing else I have ever seen," wrote Janet Linthicum, one of the hack site attendants. Many volunteers went on to careers in conservation biology.

In 1976, there was a high survival rate among the peregrines that were released in five states; by the 1980s, the Peregrine Fund was introducing the birds into more than a dozen states in the eastern US—from Maine to Georgia—and in several Rocky Mountain states. Of course the project had its critics. Some scientists were concerned that genetic purity was being lost since peregrines from Alaska were breeding with individuals from Canada, Mexico, South America, and Europe. However, as Tom points out, "There were no eastern stocks left in the US, and we used a combination of whatever we could find to replace them. However, the preponderance of the breeding stock came from North America."

Other critics feared that the introduced birds would not be able to adapt to their new environments because individuals from populations that migrated long distances were being used to build up new populations in the East, where peregrines had traditionally migrated only short distances. "But it all worked out okay," Tom said. "Some of the released birds from Arctic stock did make fall movements into South America, particularly in their first year." Others did not migrate at all, and none are known to have migrated to the Far North.

Total Restoration—Realizing a Dream

In this chapter, I have so far concentrated on the struggles and eventual success of the program in the eastern United States—because it was there that peregrines had become completely extinct, and their recovery was due entirely to captive breeding and reintroduction. But as Tom Cade pointed out, the aim of the Peregrine Fund was to restore populations of these magnificent birds to pre-DDT numbers throughout their range in the United States. In fact, as Tom stressed, most of the recovery of the peregrine in North America occurred naturally, through increased survival and increased reproduction of residual populations after DDT was banned. The Arctic peregrine falcon (*Falco peregrinus tundrius*) recovered naturally, with no captive breeding or reintroductions. And peregrines recovered well on their own in most of the Southwest and Mexico. In the West—California, Colorado, New Mexico, Utah, Idaho, Washington, Montana, and Wyoming—natural recovery was boosted by releasing captive-bred birds.

The Peregrine Fund built a second breeding facility in Colorado in 1975, and under Jerry Craig, the program reached its target of producing more than a hundred birds per year by 1985. There was also a recovery program under Richard Fyfe for southern Canada. This covered the range of the *anatum* subspecies, and only *anatum* peregrines were used for captive breeding. Taken together, Tom told me, these programs released nearly seven thousand young peregrines by hacking, fostering, and cross-fostering.

In Gratitude

In 1999, the Peregrine Fund held a celebration to mark the day that the peregrine was officially removed from the endangered species list, and more than a thousand people attended. In his address, Tom said: "My dear friends and colleagues, you and I have fought the good fight for the Peregrine, and we have won a great victory. . . . What we have accomplished together is truly phenomenal, and I believe that the recovery of the Peregrine will be recorded in the annals of conservation as a major event of the twentieth century. But, as we all know, conservation is a continual series of challenges—the fight for conservation never ends—and so I exhort you: press on to meet new challenges, for they surely await, and will always be waiting, for those who strive to keep the earth fit for life in all its many splendored forms."

Thanks to the successful reintroduction of the peregrine falcon in the major cities, Americans were given a new pastime. Shown here, an adult sitting on eggs on the 24th floor of the Union Central Building, Cincinnati, Ohio. *(Ron Austing)*

SCARLETT AND RHETT:
URBAN HEROES

One surprising development has helped to raise the peregrine's profile among the public: Falcons hacked from ledges high up on buildings in urban areas subsequently return to nest and raise their young there. The young birds were expected to move out and choose more natural places for their eyries. But there are advantages to city living, despite the occasional deaths from flying into power lines: The peregrines are safe from two of their natural enemies, the great horned owl and the golden eagle. In the Midwest, 70 percent of all nests in 2000 were in or near cities, many of them on power plants. Bridges are also favorite nesting structures. In Europe, too, wild peregrines have recently moved into cities.

Over the years, individual falcons have proved of enormous interest to people. It is now common for a video monitor to be rigged up overlooking a peregrine nest so the public can keep up with the latest developments, and Web sites have proliferated. One nest has been of particular interest. Scarlett, daughter of Sergeant Pepper and his "little Latin love," was one of the second batch of captive-bred peregrines to be hacked from an old gunnery tower in Maryland. She turned up in 1978 on the thirty-third floor of 100 Light Street, an insurance building overlooking Baltimore Harbor. The following spring, she was observed displaying and giving courtship calls to her own reflection in the window of the same building. The Peregrine Fund—which keeps a close watch over its birds—persuaded the company to install a nest tray on the windowsill; they agreed provided it matched the building's facade. So Scarlett made a scrape and laid eggs on pink Spanish granite! She soon won a large and admiring public.

Two males were released nearby, but she ignored them both, and neither stayed. Nevertheless, she laid three (obviously infertile) eggs, which were replaced (by the Peregrine Fund) with two chicks, which she successfully raised. Over the next four breeding seasons, she continued to occupy her favorite window ledge. Various males were released nearby, but none was successful until 1980, when she bonded with Rhett. Their eggs were infertile, but they raised foster chicks successfully. Unfortunately, Rhett was poisoned by strychnine in a pigeon. The released male whom Scarlett chose the following year, Ashley, recovered from a bullet wound, but then apparently died in a collision with a vehicle on the Francis Scott Key Bridge.

Meanwhile the public was following every step of Scarlett's love affairs, and there was general rejoicing when she found another young male for herself. He was named Beauregard, and the two of them raised young from her own eggs, fertile for the first time. Sadly, she then died of a massive throat infection. But the tradition of nesting on the pink Spanish granite window ledge that Scarlett had started lived on: Beauregard attracted other mates, and the public was able to follow the destiny of other peregrines.

The story of Scarlett and her beaux did much to help the public understand the plight of the peregrines. They minded when her partners were poisoned or shot. They marveled that, during the six years that Scarlett made the window ledge her headquarters, she raised eighteen foster chicks and then her own four. And they are proud that, over a twenty-two-year period since Scarlett laid her first eggs, more than sixty young have fledged successfully from their man-made eyrie at the 100 Light Street building in Baltimore.

American Burying Beetle
(Nicrophorus americanus)

The American burying beetle is but one of the millions of insects and other invertebrates that play such a major, though seldom acknowledged, role in the maintenance of habitats and ecosystems. Most people simply lump them all into the category "creepy-crawlies" or "bugs." Some, such as butterflies, are admired and loved for their beauty (though people tend to be less interested in or even repelled by their caterpillars). Others, such as spiders, are the inadvertent cause of fear—even terror. Cockroaches are loathed. Hundreds of species are persecuted for the role they play in damaging our food—such as the desert locust, which ravages crops across huge areas. And there are countless species such as mosquitoes, tsetse flies, fleas, and ticks that carry diseases that can devastate other creatures, including ourselves.

It is for these reasons that they have been attacked by farmers, gardeners, and governments. Unfortunately the weapons of choice have been chemical pesticides—and this has led to horrific damage of all too many ecosystems, either through directly killing countless life-forms in addition to the intended targets, or when poisoned insects are eaten by creatures higher up the food chain.

Yet for every species that harms us or our food, there are countless others that work away, sometimes unseen, for the good of the environment where they live. I first became aware of this when I was a small child, picking up every earthworm I found stranded on the road (as did Dr. Albert Schweitzer, by the way), and then learning about the valuable contribution they make to soil health. Millions of invertebrates provide food for species—including our own—higher up the food chain.

Lou Perrotti, coordinator for the American burying beetle for the Association of Zoos and Aquariums, is a passionate advocate for these beetles. Here he is checking a brood on Nantucket Island, Massachusetts. "Somebody needs to be out there saving these critters," he told me (note the tattoo on his forearm). (*Roger Williams Park Zoo*)

In many places people feast on termites, locusts, and beetle larvae—even I have tasted these things! Bees pollinate the vast majority of our food crops, and the current devastation of hives in North America and Europe is causing real anxiety.

And what about the American burying beetle? What role, if any, does it play in our environment? This is what I learned about when, on March 18, 2007, I met with Lou Perrotti and Jack Mulvena of the Roger Williams Park Zoo in Providence, Rhode Island. Back in 1989, they told me, biologists had realized that the American burying beetle was fast declining, and it became one of just a few insect species to be listed under the Endangered Species Act. Then in 1993, the Roger Williams Park Zoo started a breeding program for the US Fish and Wildlife Service; in 2006, this beetle became the first insect species to be assigned a Species Survival Plan. Lou is currently the coordinator for the American burying beetle for the Association of Zoos and Aquariums.

As he began talking about the beetles, it was immediately apparent that they had the perfect spokesman! He is a man passionately interested in insects and, he told me, has "loved all things creepy-crawly" since he was a child. Like so many of the other people I have talked to while gathering information for this book, Lou had parents who were understanding and supportive of his fascination with invertebrates. (And other creatures, too—they allowed him to breed boa constrictors when he was nine years old!)

While we talked, Lou became increasingly animated. "Somebody needs to be out there saving these critters [the burying beetles]," he said. And that is just what he is doing. Let me share some of what I learned from him about these remarkable beetles. Most people have no idea how fascinating they are. Certainly I hadn't.

The American burying beetle is the largest member of its genus in North America—it is sometimes called the "giant carrion beetle." Once these beetles lived in forest and scrub grassland habitats—anyplace where there was carrion of a suitable size and soil suitable for burying it—in thirty-five states throughout temperate eastern North America. But by 1920, populations in the East had largely disappeared. By 1970 populations had also disappeared from Ontario, Kentucky, Ohio, and Missouri. And during the 1980s, the beetle declined rapidly throughout the American Midwest.

Today there are only seven places where they are known to exist—Block Island (Rhode Island), a single county in eastern Oklahoma, scattered populations in Arkansas, Nebraska, South Dakota, Kansas, and a recently discovered population on a military installation in Texas. One reason for the species' precipitous decline across its historical range, in addition to habitat loss and fragmentation, is possibly connected with the extinction of the passenger pigeon and the greatly reduced number of black-footed ferrets and prairie chickens, all of which provided carrion of ideal size.

Why We Need the Burying Beetle

Let me return to the question I asked earlier—would the loss of the American burying beetle matter? The answer, stressed by Lou and Jack, is an emphatic *yes*. They feed on carrion—the flesh of dead animals. Lou calls them "nature's most efficient recyclers" because they are responsible for recycling decaying animals back into the ecosystem. This returns nutrients to the earth, which stimulates the growth of plants. And by burying carcasses underground, this industrious beetle helps keep flies and ants from reaching epidemic proportions.

Lou explained how these beetles find their meals. They can "smell" carrion from as far away as two miles, by means of sensors on their antennae. Flying noisily through the dusk, a male usually reaches the carcass he has located soon after dark. Then he—and any other males who have also discovered the feast—emits pheromones that are irresistible to females of the species. Thus, you'll likely find a number of beetles gathered around any one corpse. It seems they form pairs, and there may be a good deal of fighting until one couple claims the prize. They then cooperate to bury it. This can be hard work: A carcass the size of a blue jay will take about twelve hours to bury.

Beetle Co-Parenting

Once the carcass is safely underground, the beetles strip it of feathers or hair and then coat it with anal and oral secretions, which help to preserve the flesh that will serve as food for their young. Next, the couple consummates their pairing, and within a day the female lays the fertilized eggs in a small chamber that they have dug out close to the carcass. Here both parents wait for their eggs to hatch, which will be in two or three days. Both mother and father carry the larvae to their "larder." And then—and this really blew my mind away—the young beetles will stroke the mandibles of their parents to entice feeding, and the adults will regurgitate food for their young. How absolutely amazing—an insect species in which mother and father care for their young together!

Usually, by the time the carcass is safely underground, flies have already laid their eggs on it. These hatch quickly into hungry competitors for the young beetles. But help is close by: Riding on the bodies of the adult beetles are tiny orange mites that quickly climb onto the carcass, where they feed on fly eggs and maggots. In about two weeks, the sated beetle larvae burrow into the soil to pupate, and the parents move on. As they do so, the orange mites hop back on board. The young beetles will emerge about forty-five days later.

Lou and his team have been very successful with their captive breeding program—by the end of 2006, more than three thousand beetles had been reared and released into the wild on Nantucket Island. The captive-bred females (each paired with a genetically suitable mate) are transported to the release site in plastic containers. These are placed in an Igloo cooler, since the beetles cannot survive undue heat. A second cooler is used to transport dead quail, which the beetles will use as the carrion for their young. With a chuckle, Lou told us, "I can be traveling on a ferry during the height of tourist season and will still have room around me due to the terrible smell coming from the coolers."

At the release site, holes have been pre-dug for the beetles. The dead quail are placed into the holes with floss tied to their feet and attached to a small orange flag to assist the recovery team with finding the buried carcasses at a later date. The beetles are then released into the hole, where ideally they will realize that they have a jump start on the reproduction process! Lou said that Nantucket was chosen as a release site because, as with Block Island, there are no mammalian competitors present. After a while, though, birds such as crows and seagulls began to recognize that an orange flag represented a food source, and began to dig up the beetles' carrion, so the recovery team is now also placing a mesh screen over each brood to protect it.

Lou told me that he really enjoys teaching children about insects. We agreed that it does not take much to trigger their interest—children are naturally curious. And "creepy-crawlies," although they may elicit fear and horror, hold a real fascination for them. I told Lou I had spent hours as a child watching spiders, dragonflies, bumblebees, and the like. My son was fascinated as a little boy to watch ants as they set out in an orderly column to raid a termite nest, and returned each bearing an unfortunate victim in its mandibles. And my sister's three-year-old grandson, after watching a snail crawling over the ground, suddenly placed it on the windowpane and rushed indoors to look through the glass, clearly fascinated and curious about the mechanism that enabled the creature to glide forward, as if by magic.

Unfortunately, Lou finds it much harder to interest adults in the efforts being made to save the American burying beetle. "So often the first question," he told me, "is 'Will it eat my garden?'" If only people would take the time to listen, retain the curiosity and wonder of childhood, how much richer their lives would be. Certainly during my short early-morning meeting with Lou and Jack, I had been transported to a different and utterly fascinating world, where giant insects nurture their young and tiny mites, in exchange for a free meal and a ride to the restaurant, rid their benefactors of their competitors.

After our visit, Lou sent me a beautiful print of an American burying beetle, its orange and black colors vivid and glowing. It is propped against the wall as I write, reminding me of all the magic of the natural world.

Crested Ibis
(Nipponia nippon)

I first learned about the Chinese scientist Dr. Yongmei Xi, and her remarkably successful efforts to save the crested ibis from extinction, from George Archibald of the International Crane Foundation (ICF). He said these birds were among his favorites, and he even sent me photographs to show me how beautiful they are. Amazingly, two weeks after my conversation with George I was able to meet with Yongmei Xi herself while I was in Shanghai in 2007—what a privilege! While we drove from one locale to the next (there was no other time), Yongmei and I talked about these special birds and her love for them.

At one time, the crested ibis was plentiful in the wetlands of Japan, China, Korea, and Siberia. By 1930, however, there were very few left: They had been relentlessly hunted, especially for their glorious feathers, but also because women believed that eating ibis would help to restore their strength after childbirth. By the end of World War II, in 1945, it was ascertained that the remaining populations had been almost exterminated throughout their range as a result of hunting, pesticide use, and habitat loss. Particularly disastrous was the draining, during the winter, of previously wet paddy fields to control the spread of snail-borne disease to humans.

It is interesting that ibis, over time, seem to have evolved a dependence on humans—they need the rice paddy habitat. They roost and breed in trees on the higher slopes, and are most at ease when there are humans living near the trees they select for nesting.

By 1978, the crested ibis was extinct in Korea. (George Archibald made a heroic effort to catch the last four—for captive breeding—in their wintering grounds in the Korean Demilitarized Zone. But his mission failed.)

Yongmei Xi's passion and determination helped prevent the extinction of this beautiful bird. She had this snapshot taken so she would have a picture of a crested ibis to take with her on a long journey. *(Yongmei Xi)*

Black-Footed Ferret (*Mustela nigripes*). Once lost in the wild, but now saved and restored in selected areas throughout their original range of the great North American prairies. (© *Thomas D. Mangelsen*)

Mala or Rufous Hare Wallaby (*Lagorchestes hirsutus*). One of the captive-bred malas released in Alice Springs Desert Park, Australia. *(Peter Nunn)*

California Condor
(*Gymnogyps californianus*).
Saved in the nick of time through
captive breeding. "XEWE"
was an adult male mentor
for captive-bred condors
in Baja California.
(*Mike Wallace*)

Red Wolf (*Canis rufus*). Once completely extinct in the wild, and now being restored. Shown here, a father grooming one of his pups in North Carolina. (*Greg Koch*)

Tahki or Przewalski's Horse (*Equus ferus przewalskii, Equus przewalskii or Equus caballus przewalskii*—classification is debated). This Mongolian native horse became extinct in the wild in 1968. Thankfully, a few still existed in European zoos, and in 1994 they were restored to their native homeland in Mongolia's Hustai National Park. *(Christopher A. Myers)*

Golden Lion Tamarin (*Leontopithecus rosalia*). One of the many golden lion tamarins now freely roaming their ancestral home in the Brazilian rain forest. *(Mehgan Murphy, Smithsonian National Zoo)*

American Crocodile (*Crocodylus acutuse*). A unique photo of an adult feeding in the Florida Everglades. The recent comeback of this shy crocodile is an endangered species success story, but the animal's future depends on the restoration of the Everglades. (*Joseph A. Wasilewski*)

Peregrine Falcon (*Falco peregrinus*). The peregrine falcon became completely extinct in the Eastern United States, mostly because of widespread use of DDT. Their resounding US recovery in this area was entirely due to captive breeding. Shown here, a mother with chicks, feeding in the wild, Nunavut, Canada. (© *Thomas D. Mangelsen*)

American Burying Beetle (*Nicrophorus americanus*). Known as "nature's most efficient recycler," the American Burying Beetle almost became extinct in the 1900s. If we were to lose this crucial carrion beetle, ants and flies could reach epidemic proportions. Shown here, a wild beetle on Nantucket Island, Massachusetts. (*Roger Williams Park Zoo*)

Crested Ibis (*Nipponia Nippo*). Once there were only nine of these stunning birds left in the world. Now almost a thousand exist in China. *(Bjorn Anderson)*

Whooping Crane (*Grus Americana*). Living free in the breeding grounds of Saskatchewan, Canada. Herculean efforts, by countless dedicated men and women, have saved the whooping crane from extinction. (© *Thomas D. Mangelsen*)

Angonoka or Ploughshare Tortoise (*Geochelone yniphora*). These two males are jousting, competing for dominance; the victor will mate the female of his choice. The captive breeding of this highly endangered tortoise from Madagascar began with only eight individuals, five of whom were males. (You can see the "plough" from which the species name is derived under the loser's chin.) (*Don Reid*)

Vancouver Island Marmot (*Marmota vancouverensis*). Onslo, a captive-born marmot from the Toronto Zoo, now looking toward a brighter future at Haley Lake, Vancouver Island. His mate Haida was the first captive-born marmot to breed successfully in the wild after her release in 2004. *(Andrew Bryant)*.

Sumatran Rhino (*Dicerorhinus sumatrensi*). Emi stands with her calf Suci, the second Sumatran rhino born at the Cincinnati Zoo and Botanical Garden. As with the California condor, captive breeding for this rare rhino was initially controversial, but perseverance is paying off. Andalas, the first Sumatran rhino born in captivity, has been returned to his native Indonesian home and now lives in a protected area in Sumatra. *(David Jenike)*

Gray Wolf (*Canis lupus*). The return of the gray wolf to its native habitat of Yellowstone National Park heads the official "World's Top Ten Conservation Programs" list. Although wolves remain controversial in the American West, their comeback is proof that people and predators can live together. (© *Thomas D. Mangelsen*)

Iberian Lynx (*Lynx pardinus*). What a tragedy if Spain loses her highly endangered Iberian lynx. Fortunately, captive breeding is ahead of projections, and the wild population is slowly growing. We just have to secure their habitat. Shown here, Saliego, with her two cubs Camarina and Castañuela, born in 2006. (*Héctor Garrido*)

Bactrian Camel (*Camelus bactrianus ferus*). More endangered than the giant panda, this unique wild camel from China and Mongolia is being saved through captive breeding and habitat protection. This is the only photograph ever taken of a wild Bactrian camel and her newborn calf. (The mother wandered by herself into the unforgiving Gobe Desert to give birth; the calf is not a day old.) *(John Hare)*

Pygmy Hog (Porcula salvania). A successful captive breeding was launched in 1996 by capturing six pygmy hogs from the last surviving population in Manas National Park in India. From these six original individuals, there are now about eighty pygmy hogs. *(Goutam Narayan)*

Giant Panda (*Ailuropoda melanoleuca*). Su Lin—whose name means "A little bit of something very cute"—was born on August 2, 2005, at the San Diego Zoo. Eventually she will be moved to a captive breeding nature reserve in China. One of the biggest challenges faced by giant pandas is the lack of a suitable habitat. *(Ken Bohn)*

Northern Bald Ibis or Waldrapp (*Geronticus eremire*). Speedy is one of the first northern bald ibises from Austria to successfully learn a new migration route by following an ultralight aircraft. The birds' ability to migrate to Italy's warmer climate in the winter is key to their survival in Europe. (*Markus Unsöld*)

Columbia Basin Pygmy Rabbit (*Brachylagus idahoensis*). These highly endangered and adorable rabbits are being captive bred to insure their survival. This is the first (and only known so far) offspring of reintroduced pygmy rabbits in Eastern Washington. *(Len Zeoli)*

Attwater's Prairie Chicken (*Tympanuchus cupido attwateri*). At one time these birds ranged across some six million acres of American prairie—but less than one percent of their original prairie remains. Shown here, a male "booming" display at National Wildlife Refuge, Eagle Lake, Texas. *(Grady Allen)*

Asian or Oriental White-Backed Vulture (*Gyps bengalensis*). The Asian vulture population fell by more than 97 percent in less than a decade—one of the steepest declines experienced by any bird species. Efforts to save and restore this species are underway through captive breeding and widespread public education efforts. Shown here, a wild trio sunning themselves in the foothills of the Himalayas at Corbett National Park, India. *(Nanak C. Dhingra)*

In 1981, the last five individuals remaining in Japan were captured and taken to a breeding center—but they did not breed.

China Searches for the Last Ibis

Meanwhile there was growing concern for the fate of the crested ibis in China. Dr. Liu Yen-zhou, of the Institute of Zoology in Beijing, organized surveys to look for them in central China, but for the first three years the team saw no sign of the crested ibis. Then in 1981, they spotted a group of seven in the Tsinling Mountains, not far from the ancient capital Xian.

The Ministry of Forestry at once agreed to provide protection for these precious individuals—the last of their species. Farmers were paid *not* to apply toxic chemicals to wet rice paddies, and as a result the habitat gradually improved. At the same time, they devised some innovative techniques to give the birds as much help as possible. By wrapping the trunks of nest trees with smooth plastic materials, predation from snakes was reduced. By putting nets under the nests, weak chicks evicted by stronger siblings could be either put back and given a second chance, or—if they were very weak (sometimes a chick was evicted a second time)—cared for in captivity. These birds would subsequently become part of a captive breeding program.

Breakthrough in Captivity

As a result of all these measures, the wild population began to increase. But very slowly. Yongmei first studied the crested ibis in 1988. She told me that a pair of crested ibis in the wild has only one clutch each year; an average of two chicks survive. In captivity, however, Yongmei found that a pair can have two to three clutches from which an average of seven chicks survive. And so, in 1990, it was decided to start a breeding program, and by 2006 there were a total of four centers in China.

Yongmei, meanwhile, was becoming more and more familiar with these beautiful birds, and partly no doubt because of her empathy with them, she and her team were very successful in their breeding program. She tried to supply the captives, so far as possible, with a diet that included food items eaten in the wild—such as loach, a common small fish, which they find in the paddy fields. She told me how excited she had been when, for the first time, a pair of ibis who had been born in captivity managed to successfully raise their own chicks. Before this, the parents had sometimes destroyed their eggs or killed their chicks, and she had come to believe that this was because the enclosures were not suitable.

And so, in 2000, she constructed a large cage of green nylon on the slope of a mountain. It was surrounded by trees, and there were real trees growing inside. The breeding success in this enclosure provided the first evidence that parents could take care of their chicks in captivity if the conditions were to their liking.

While observing the captive birds, Yongmei noticed how they interacted with various wild birds who were attracted by the food. "When the wild birds land on the wire of the cage roof," she told me, "they call out to the captives, who return their calls." She believes that the wild birds envy the captives their plentiful supply of food—but she does not believe that the captives are satisfied with their mainly pellet diet. She thinks they long to fly up and away with the visitors when they leave.

Yongmei also told me about two of her young captive ibis, who were sent to the emperor of Japan as a gift in 1999. Understanding them so well, she felt sure that they would be lonely in their new surroundings. Their natural food—loach—and the pair's original food container were sent with them to Japan. Twigs were provided for the pair in the breeding season. They began to lay eggs, and eventually one of the eggs hatched. It was a male. The next year another female bird was sent to Japan from China. Based on these three founder birds, a new ibis breeding program was established. In 2008, I was told, there were 107 captive crested ibis in Japan.

Back to the Wild

As of 2008, there were also about a thousand ibis in China—five hundred in the wild, and another five hundred in captivity—and there are plans to introduce some of the captives into the wild. A major effort is under way to restore their habitat in the Hanzhong Basin. The use of agricultural pesticides is strictly controlled, and a series of handmade reservoirs linked to a network of rivers will improve things for the birds, and for the rice farmers. Also, some grassland will be flooded. There is an education program in which people in ninety-one villages in the area are given information about the crested ibis and its habits.

Perhaps, one day, I shall be able to see this glorious bird in the wild. I am so grateful to George for sending me a beautiful photograph of a crested ibis in flight, and most of all for introducing me to Yongmei so that I could hear the remarkable story I have shared here from her own lips.

Whooping Crane
(Grus americana)

There is something almost mystical about cranes. They are an ancient genus, and their voices, loud and wild, seem like echoes of the past. They are also elegant birds, with long legs, long necks, and long sharp beaks, all suiting them for the grasslands and wetlands where they forage. There are many species of cranes in the world today: Almost all of them are endangered.

This chapter describes the Herculean efforts, by countless dedicated men and women, to save the whooping crane from extinction. They are the only cranes native to North America. Standing between four and five feet high, they are magnificent, with snowy white plumage except for a brilliant red cap on the top of their head, black facial markings, and black primaries clearly visible in flight. With their long spear-like beaks and fierce golden eyes, they can be formidable when protecting their young.

When Europeans first arrived in North America, it's estimated that whooping cranes numbered at least ten thousand. They wintered in the highlands of central Mexico and on the Gulf Coast of Texas and Louisiana, as well as the southeast Atlantic seaboard, including Delaware and the Chesapeake Bay. Their breeding grounds were many—throughout the central prairies of the United States and well into central Alberta, Canada. But by the end of the nineteenth century, migratory whooping cranes were no longer breeding anywhere in the US. And by 1930, they were no longer breeding on the Alberta prairies. In fact, no one knew where the last migrating birds were breeding, except that it was somewhere in Canada.

One nonmigratory flock of whooping cranes in Louisiana continued

George Archibald dancing with Gee Whiz, the only offspring of Tex, a famous female whooping crane that George patiently "courted" and bonded with in order to bring her into reproductive condition. *(David Thompson / ICF)*

nesting there through the 1930s, but in 1940, when only thirteen birds remained, a hurricane scattered this remnant group, and though six survived, they were doomed. And by this time, fewer than thirty of the migrating whooping cranes were arriving in the fall in Texas (in the Aransas National Wildlife Refuge) from their unknown breeding grounds in the Canadian North. The days of the whooping crane seemed numbered, and most people felt that nothing could be done to save them.

But some were determined to try. Three organizations—the US Fish and Wildlife Service; its Canadian equivalent, the Canadian Wildlife Service; and the Audubon Society—collaborated in a desperate attempt to prevent the species from becoming extinct. First they needed to find out more about them. Most of what they learned was depressing: Cranes were being shot by hunters, or farmers who resented them as potentially destructive to their crops—one publicly vowed to "shoot the pesky things on sight." In 1953, only twenty-one cranes arrived in Texas.

As a last resort, the wildlife organizations launched an awareness campaign. The Whooping Crane Conservation Association got involved and helped spread the word. They informed people along the migration route—so far as it was known—about the cranes, their history, and the current dire situation. And appealed for their help. It worked, and the shooting stopped. Meanwhile private citizens in the organization were lobbying to get governments to take action and provide the cranes with better legal protection.

In 1954, there was a breakthrough: Canadian Forestry Superintendent G. M. Wilson and his helicopter pilot, Don Landells, spotted two white birds with a cinnamon-colored chick in the boreal marshes and ponds of the remote Wood Buffalo National Park in northern Canada. They had discovered the last breeding grounds of the whooping cranes! The birds were migrating a staggering twenty-four hundred miles twice each year from northern Canada to Texas and back again.

Gradually, as a result of protective measures and the awareness campaign along the migration route, the tiny flock increased. In 1964, forty-two birds arrived in Texas, and the following year the number was even higher. But the situation was fragile. And so, in 1966, the Canadian Wildlife Service (CWS) and USFWS finally agreed to collaborate to establish a captive breeding program. Not everyone thought this was a good idea, but the two national wildlife agencies moved forward with their plan.

Meet Ernie Kuyt, Egg Thief!

Ernie Kuyt, whom I phoned on the recommendation of my friend Tom Mangelsen, was one of the first people brought in to work on the breeding scheme. During a long conversation, Ernie said that he had become involved with the whooping cranes by accident. CWS had needed a field biologist to help find the nests and safely transport surplus eggs for a captive breeding colony, and Ernie was the only one available.

A plan was formulated: Cranes normally lay two eggs, but typically rear only one chick—and often only one egg is actually viable. So whenever they found a nest with more than one egg, Ernie would test them. "It was crane biologist Rod Drewien," said Ernie, "who taught me how to test the viability of eggs at the nest by simply floating them briefly in lukewarm water." (I am familiar with that process—I tested every hen's

Ernie Kuyt collecting one of the two eggs from a wild crane nest in Wood Buffalo National Park, Canada. He carried every precious egg—for captive breeding— in one of his thick wool socks. *(Ernie Kuyt)*

egg before buying it in the early days in Tanzania!) If both eggs were good, Ernie took one of them. If a nest had only bad eggs, he would remove them and replace them with one of the good eggs he had collected from another nest. All the excess eggs he collected were sent to hatch at the Patuxent Wildlife Research Center in Maryland to start the captive breeding population.

Ernie told me about the time he left the base on June 2, 1967, to collect the very first egg. "The Americans had designed a special Styrofoam case to carry each precious egg from nest to base," he said. "It was only as the helicopter was preparing to land that I realized I had forgotten the box!" They could not return because that would upset the time schedule—and the budget. Yet he remembered only too well the ominous memo from HQ that had warned: "You will agree that no slip-up is possible!" Much was at stake, and all eyes were on Ernie and his team.

Fortunately, knowing he would get his feet wet slogging through the marshes, Ernie had brought heavy wool socks along. Carefully, he lowered one of the two good eggs into a sock until it nestled gently in the toe. Carrying the sock by the cuff, Ernie transported the egg safely back to the waiting helicopter. "It worked so well that the fancy egg-case was never used," Ernie told me. "During my twenty-five years of crane work, I safely transported over four hundred eggs without damaging a single one—using thick wool socks!"

Stories from the Field

Ernie told me a story about one pair of cranes, known as the Hippo Lake pair, which had built their nest near a lake shaped in the form of a hippopotamus. On one of his aerial surveys, Ernie noticed that their nest was empty. Several days later, he saw a single egg. But two days later, "the egg was gone, though one of the birds was still attending the nest." Eleven days later, on the day of an egg pickup, Ernie flew past the Hippo Lake nest one more time. The crane was incubating on the nest—but when she stood up, Ernie saw that the nest was still empty.

"The adult birds had been on the empty nest for almost two weeks! Were they telling us something?" When the biologists landed the chopper, Ernie put an egg in the nest that they had just collected from another nest. The Hippo Lake pair hatched that foster egg, and Ernie had the happy task of banding the chick before it fledged.

Whenever Ernie was on the ground, an aircraft circled overhead, monitoring the scene so they could warn him of nearby bears or moose. Once, as he approached a nest, the Cessna made a shallow dive over-head—their code for danger—and he saw a black bear moving toward him. Luckily, it was not fully grown—probably two or three years old. "I picked up a dry tamarack stick and began beating it against a tree, at the same time yelling at the top of my voice," said Ernie. The bear, about thirty yards away, looked at him, then turned and ran off. The eggs in the nearby nest were so close to hatching that the distinctive peeping sounds of a chick were clearly audible. If Ernie had not driven it away, the bear would almost certainly have found and raided the nest.

Tracking the Migration

Ernie not only collected eggs, but also followed the cranes in a Cessna 206 when they migrated, radio-tracking them and collecting valuable new information. One fall he invited Tom Mangelsen to join him, to document the journey with film and stills, and to keep track of the cranes visually while Ernie was busy plotting the route, and the pilot was concentrating on flying the plane.

Migrating cranes use the thermals to spiral up and then glide seem-ingly effortlessly on their great wings. "On days of bad weather with head-winds, they would fly little or not at all," Tom told me, "but on good days they could cover four hundred miles or more." Fortunately the whooping cranes, with their white plumage and huge wingspan, were relatively easy to see. "We were able to keep visual contact with them nearly 50 percent of the time," said Tom, "and we could pick up the radio signals transmit-ted by the birds within a radius of twenty-five to a hundred miles.

"Watching the cranes flying with such grace against a limitless sky and endless landscape," Tom told me, "was the most inspiring event of my life."

Ernie felt the same. He told me, "The ability and opportunity to migrate with the cranes . . . has been the highlight of my twenty-five-year study."

One Flock Is Too Fragile

While Ernie and others were protecting the Wood Buffalo/Aransas flock, crane biologists and conservationists on the US and Canadian Whooping

Crane Recovery Teams were planning other initiatives. The single remaining wild flock was just too fragile: If disease or disaster struck, it could be annihilated just as the Louisiana flock had been.

The first plan involved placing whooper eggs in the nests of sandhill cranes nesting in Idaho. This initiative failed because, while the fostered chicks did indeed follow the sandhills to New Mexico, as hoped, they never courted and mated with their own species. A young crane, like many bird species, becomes imprinted on its parents soon after hatching, and if at this critical time a bird of the same species is not available, the chick will become imprinted on almost any moving object. Unfortunately, these whoopers were imprinted on the sandhills and courted the sandhills when they reached maturity.

Meanwhile a number of experts, including George Archibald, cofounder of the International Crane Foundation, believed they should try to establish a nonmigratory flock in Florida, in the vast area of Kissimmee. In 1993, the first group of captive-bred crane chicks arrived there for release into the wild. And after that, every year until 2005, further chicks were sent to boost numbers. These birds formed pair bonds, established territories, and built nests just like wild birds. But there were many problems—especially predation by bobcats. In 2005, despite all the hard work and the great hopes, it was decided to discontinue the release of captive-born chicks, and the outlook for the future of the few surviving Florida cranes is bleak.

Cranes, Men, and Their Flying Machines

Although things were going well with the one migratory flock, two costly attempts to establish new flocks had failed. There was still a need to establish a new migratory flock—and an innovative idea was being suggested. What if it were possible to teach young cranes to follow an ultralight aircraft? At a conference in California, I heard a talk about this by Bill Lishman, an inspired and passionate naturalist. Eventually he had partnered with Joe Duff, an ex-businessman, and working with nonendangered Canada geese, the two men gradually perfected the technique—which was introduced to the public in the popular movie *Fly Away Home*.

During the late 1990s, after working with sandhill cranes, Bill and Joe presented their results at the annual Canadian/US Whooping Crane

Recovery Team meetings, hoping to convince the team to use this method for whooping cranes—but it took five years before the plan was approved (many felt that Bill and Joe were only interested in making another movie!). Operation Migration was born in 1999 with the goal of teaching young captive-born whooping cranes to fly from Wisconsin to Florida.

Operation Migration

In 2006, I received an invitation from Joe—would I like to experience, firsthand, the training of the whooping cranes? Fly in an ultralight? My schedule was packed, but this was something I could not refuse and I freed up two days during my US/Canada fall tour. Two days I shall never forget.

Joe Duff and operations manager Liz Condie met me at the Madison airport in Wisconsin. It rained, quietly, throughout the one-hour drive to the trailer camp at the Necedah National Wildlife Refuge. And every time I woke during the night, I heard rain pattering on the trailer's metal roof. It seemed unlikely the weather conditions would allow us to fly in the morning.

Wearing my crane suit before flying in an ultralight with Joe. Operation Migration trains captive-bred cranes to migrate from Wisconsin to Florida by following an ultralight "parent." (© www.operationmigration.org)

Indeed, the morning weather was unsuitable, so instead I met more of the team and learned about the program. Earlier in the year, eighteen cranes, about forty-five days old, had arrived from Patuxent Wildlife Research Center. In order to prevent these whooper chicks from imprinting on their human foster parents, those who rear and train them for release wear white gown-like costumes, black rubber boots, and helmets with visors that hide their eyes. They carry tape recorders that play the brood calls of parent cranes and the sound of the ultralight that the chicks will learn to follow. In one hand, the handler holds a puppet looking like an adult crane's head and neck, complete with gold eyes, long dark bill, and distinctive red crown. The sleeve of the costume, which covers the hand and arm, blends into the long white neck of the puppet (a metal tube covered by white cloth). There is grain in the "neck" that can be released through a hole as the puppet pecks the ground.

In Necedah, during the summer months preceding the fall migration, the Operation Migration crew of pilots, biologist, veterinarian, and interns continues the education of the young birds that was started in earliest chick-hood at Patuxent.

That same morning, I also visited the adolescent cranes in their closed-in pen, half of which is in shallow water. They were beautiful golden-and-white feathered youngsters. I put on one of the crane suits, borrowed a crane puppet head, and followed Joe and two other pilots, Brooke and Chris, to the pen, stepping through a pan of disinfectant on the way. I could not believe I was actually taking part in this extraordinary and inspirational project, and felt tears stinging my eyes. Once we were within earshot of the cranes, there was no more talking.

The young cranes, who had learned to live together as flock members, were as tall as adults but still wore the white-and-golden plumage of adolescence. Their long, black-tipped wings had been strengthened by their daily training flights, and they were almost ready to set out on their twelve-hundred-mile journey to Florida. They were very curious and investigated everything that caught their fancy, gently probing with their beaks. From time to time, one of my fellow human cranes approached me with a grape; I opened the beak of my puppet with a lever, took hold of the fruit, and offered it to one of the cranes. They love grapes.

There was a sense of mystery, the feeling that I was in the presence of ancient bird wisdom, and connected with an *other-than-self* life force. My humanity was diminished. And then one of the birds pulled at the tip of my "wing," while a second prodded my boots and a third had a go at the felt of the puppet head so that I had to move it away and engage him—or her—beak-to-beak. I had no sense of the passing of time, and much too soon we had to leave them.

Flying with Cranes

When I looked outside at six the next morning, the sky was clear, and there was almost no wind. A perfect day for flying! At the hangar, I donned my white crane costume; then came earphones, and finally the helmet. The pilots wheeled the ultralights out of their hangar, and I climbed into the tiny passenger seat in the space behind Joe. After we belted up, he attached my headphones to the system so I could hear him, pulled the cord to start the engine, taxied to the runway—and we took off.

The golden and pale blue morning air was all around, rushing past us, exhilarating. For the first time ever I felt that I was truly flying, part of the air, and the clouds, and the sky. Spread out over the waking landscape, the other three ultralights flew toward the landing strip adjoining the cranes' pen. There we all touched down and the cranes were let out to join their strangely assorted parent figures—disguised humans and unlikely flying machines! One of the four pilots, Chris, taxied carefully through the eighteen cranes and about seven of them followed, running after the plane; when he took off, so did they. Up they flew, parent ultralight and its little following. The remaining youngsters on the ground milled around pilot Brooke, making it very hard for him to take off, but he made it with all but one of them flying up after him. He flew in a big circle and swooped back past the remaining crane, which then decided to follow.

Soon we were all in the air. Because of my extra weight, Joe could not reduce his speed sufficiently to have the cranes actually follow us—but we were often very close to them. The pilots communicate with one another, so they can turn to pick up a crane that has flown off on its own, or know when two or three more join their little flock. One of the cranes absolutely got the hang of gliding in the slipstream of the ultralight he was following, and scarcely had to flap his wings.

It is hard for me to describe the emotions that went through me as I sat there behind Joe. I felt so much part of the whole scene, flying in that frail little machine above the wildlife refuge, the other ultralights like huge birds, each with its cranes strung out behind, the glory of the morning with its after-rain freshness and rising sun and golden clouds. The reflection of plane and cranes shone in the calm surface of the water below. I developed a new feeling for the cranes themselves on an almost spiritual level of connectedness.

I wanted to go on flying forever, suspended between heaven and earth with those exquisite young whooping cranes. If only the engine had been silent, the experience would have been unearthly and I could have believed myself a bird.

I called Joe regularly during the long weeks of the migration—it was shocking how many flying days were lost because of bad weather. At last came the news I had been waiting for: All the birds had made it to Florida. After a journey of twelve hundred miles, all were safely in their spacious new winter home at the Chassahowitzka National Wildlife Refuge. The human team could return to homes and families. And for the cranes, life would finally settle down. Experienced handlers would pen them at night, releasing them each morning to explore their new habitat. The pen, built in a large pond, had two purposes: to keep the chicks safe from nighttime predators, and to continue to teach them to roost in water at night.

And then, a few months later, Joe called me again, this time with devastating news. All but one of those glorious crane beings were dead, killed in their pen by lightning during a freak storm that also killed twenty people. Yet setbacks like this must be endured, time and again, in the fight to rescue animals pushed, by us, to the brink of extinction. Joe and the rest of the Operation Migration crew would carry on.

There was good news for that same year: In the summer of 2006, at least six pairs of cranes nested and laid eggs in Necedah—and although only one chick fledged, it followed its human-trained parents to Florida. The following spring (2007) the two adults—known as the First Family—once again nested and laid an egg in Necedah.

Meeting the Eggs—and Other Birds—
at Patuxent Wildlife Center

On a glorious spring day, five months after my flight in the ultralight with Joe, I visited the whooping crane breeding program at Patuxent Wildlife Research Center in Maryland, where two-thirds of all whoopers so far released into the wild were raised. The director of the crane program, John French, along with several members of his team, was there to greet me and explain what was going on. Currently Patuxent is in charge of rearing and training all the chicks destined for Operation Migration. This training is carried out by a team of scientists, veterinarians, support staff, and the crane handlers who directly care for the birds. Many of those working at Patuxent have been there ten to twenty years, giving the crane project consistency and stability.

The eggs come from Patuxent's own breeding birds, and are also sent from ICF and other facilities. At the time I visited, there were forty-five eggs in various stages of incubation and "chick season," as the Patuxent crew calls it, was in full swing. One egg was actually hatching while I was there, and I went to visit it. Chicks must not hear human voices even in their eggs; as mentioned, they hear recordings of crane brood calls and the sound of the ultralight plane from the earliest age. These recordings, they told me, are played at least four times a day during the entire hatching process.

As we approached the hatching egg, we could hear the desperate-sounding peeping of the chick as he struggled to break through the shell, and every so often a small beak appeared through the little square hole he had already chiseled. I longed to help, but the initial fight to emerge is, said John, critical to the chick's survival. Chicks that cannot hatch on their own are often weak; in the wild they probably would not make it. Those that break free on their own are usually robust, as if the difficult two-day process also encourages qualities of persistence and determination—very important for a bird destined for a demanding existence in the wild. (We named that struggling chick Addison after a friend of mine who has made generous donations to Operation Migration.)

Next, I again donned a crane suit and accompanied a two-week-old chick on his daily walk to the wetlands area, along with crane handlers Kathleen (Kathy) O'Malley and Dan Sprague. This regular exercise is necessary to strengthen their rapidly growing legs. It also acclimatizes the chick to a wetlands environment where it learns to hunt, following the example of the human-wielded puppet head as it probes, crane-like, the ground and water.

On the way back, the chick, along with his "parent," followed a noisy ultralight around a small circular track. As he grows older, he will learn to follow the plane when it is driven around the track by his handler. At this time, the regular puppet head is exchanged for one with an extremely long neck (known as a robo-crane) so that the handler can continue to interact with his chick even when sitting in the plane. A robo-crane, like the puppet I used at Necedah, can dispense mealworm "treats" to the always hungry chick each time the handler pulls a trigger—it is important to reward them frequently for following the plane. Chicks start this daily training as early as five days of age. By the time they're sent to Joe and the Operation Migration team in Wisconsin, they have been following the plane for weeks on the ground and are ready to start flying lessons.

Disease, Heartbreak, and Continued Determination

Four months after my visit to Patuxent, I learned that out of the forty-five eggs that were there at that time, only seventeen chicks would be available for shipping by private jet to Operation Migration in Wisconsin. Kathy explained that a variety of diseases and genetic problems—such as scoliosis, heart issues, and weak legs—were responsible for the loss of chicks. She has been involved with the whooping crane breeding program since 1984 and has raised more than three hundred whooper chicks, a world record! She definitely has a flair for this work—during her first year in charge, the survival rate went from less than 50 percent to 97 percent.

She told me that she has spent many nights struggling to save whoopers, and has had to work around the clock with veterinarians for weeks at a time. Once a toxic mold grew in the feed and 90 percent of the birds (sandhills and whoopers) became sick. "We had to tube-feed almost all the birds to save them," Kathy recalled. "We worked for six weeks without a single day off . . . That was a terrible time. But we got through it."

It was Joe who told me that his dream of leading a much bigger flock that autumn was not to be. "But at least we have seventeen birds to train—and there were not many more in the whole world when the first efforts to save whooping cranes were made." Addison, he assured me, was doing really well—"strong and feisty."

A Visit to the Original Flock in Texas

Meanwhile the wild Aransas/Wood Buffalo flock, which provided those first eggs for the first captive-bred chicks at Patuxent, has steadily increased. In the fall of 2006, 237 birds returned from Canada to Aransas in Texas, with 45 fledged chicks including a new record—*seven* "twins" (meaning both eggs hatched from seven two-egg clutches). And the following year, 266 wild whooping cranes wintered in the refuge.

The Aransas National Wildlife Refuge was founded by President Franklin D. Roosevelt in 1937 to protect the migratory and other birds that find a rich food source—blue crabs and other aquatic creatures—in the brackish pools of the marshland habitat. We would not have a story to tell if this land had not been protected back then. Unfortunately the wetlands along the coast of Texas have become increasingly degraded due to human population pressures, heavy commercial shipping, and the introduction of exotic species. And fifteen hundred acres of the refuge was lost when a channel was dredged for the Intracoastal Waterway that cut right through the six thousand acres of marshland.

By the start of the new millennium, it was estimated that some 20 percent of the original refuge had been lost. Finally, it was decided that something must be done. A major effort to protect and restore the marshlands is now under way: The banks along the waterway have been lined with heavy matting that completely stops the erosion of the salt marsh. New levees have been built, and material dredged from the channel has been piled up on the inside of the barrier and seeded with marshland plants. It is hoped that the cranes will eventually move into this man-made habitat.

I was in Aransas in 2002 to help celebrate the hundredth anniversary of the entire National Wildlife Refuge System. My visit had been arranged by ConocoPhillips—for many years, Conoco had contributed funds to the preservation of the marshland. At the dinner Tom Stehn, whooping crane coordinator with the US Fish and Wildlife Service, Aransas, presented me with a treasured feather from the wing of a whooping crane (with all its government permissions for ownership!). But before that, there had been time to go out in the research boat. As we moved slowly along the waterways, a roseate spoonbill flew past, the pink of its wings illuminated by the setting sun. And then, filling the air with magic, came the call of the whooping crane. There they were, a pair standing tall and straight, then bowing their heads to search for the blue crabs and frogs of the wetlands. We saw two more pairs before the dusk closed in and we had to head back. We did not try to get close—it was enough to know that they were there, still returning to their old ancestral winter feeding grounds. And one last time, we heard the wild call of a whooping crane sounding over the darkening wetlands.

This picture is vivid in my mind as I sit, thinking back over the past few years. Despite everything, against all odds, these ancient birds have survived, and it is thanks to the imagination, dedication, and sheer determination of the people I have met during this journey of discovery, and all those I have not. People who have devoted their lives to ensuring that the whooping crane shall not vanish from the marshlands and prairies, rivers and skies, of North America.

THE ROMANCE OF GEORGE AND TEX

George Archibald has devoted his whole life to cranes of all species. He has played a role in the conservation of the whooping crane—and not only in conventional ways. The story of his courtship with a whooping crane named Tex is enchanting.

Tex, who hatched at the San Antonio Zoo in 1966, was hand-raised and imprinted on humans. She was a rare and valuable bird carrying unique genes, and it was important that she reproduce—but a decade of introductions to suitable male cranes failed. Tex preferred male Caucasians. George knew that hand-raised cranes will sometimes lay eggs if they form a close bond with a human—so he volunteered to "court" Tex.

In the summer of 1976 Tex arrived at the International Crane Foundation, where a shelter had been built for the unconventional couple. Tex's side was equipped with two buckets—one for fresh water and one for nutritious pellets. George's side had a cot, a desk, and a typewriter.

Most of the day, Tex stood nearby and watched George, but sometimes she led him outside.

Cranes have a remarkable courtship dance that includes bowing, jumping, running, and tossing objects into the air. To strengthen their bond, George agreed to join Tex in this elaborate performance many times daily during the early months of their relationship.

And it worked. The following spring, Tex laid her first egg. Unfortunately, although it was artificially inseminated, this egg was infertile. So their courtship dancing continued. The next spring she again laid one egg, but to George's intense disappointment the chick died while hatching. And for the next three years George was working in China, so others danced with Tex. But she never laid for them.

"In the spring of 1982, I made an all-out effort with Tex," George told me. For six weeks he spent every hour with her, from dawn until dusk, seven days a week. Once more, she laid one egg. And this time the chick hatched. He was named Gee Whiz.

Three weeks later, as George was about to appear on *The Tonight Show with Johnny Carson,* he heard that Tex had been killed by raccoons. He went onto the show anyway, and after sharing his courtship story with twenty-two million people, he broke the sad news.

"The studio audience gasped, and that ripple of anguish was felt nationally," he said. "Through her dance and death, I think, Tex made a great contribution to public awareness about the plight of endangered species."

Gee Whiz prospered and eventually paired with a female whooper. Many of his offspring have been introduced back into the wild, and the genes from Tex are alive and well in both captive and wild populations of whooping cranes.

Angonoka or Ploughshare Tortoise
(Geochelone yniphora)

My friend Alison Jolly, a renowned primatologist and author, first told me about the angonoka or ploughshare tortoise, which lives in a remote area of northwestern Madagascar known as the Soalala peninsula. It was called the ploughshare (or plowshare) tortoise because part of the lower shell sticks out between the front legs like a plow.

"They are marvelously funny animals," Alison told me. "The males joust with the long 'plowshare' spur on their lower shell that sticks forward under their chins. The goal is to tip one's rival over on his back. They are big, like soccer balls. The one on his back rocks wildly as he struggles for a foothold to turn over again." Although for the losing male it is, without doubt, a very undignified situation and not funny at all!

These tortoises live within a six-hundred-square-mile area of bamboo scrub forest and open savanna. Without the dedication of a group of conservationists, it seems almost certain that they would have slipped over the brink into the abyss of extinction. The tortoises were not hunted for food, but irresponsible dealers were taking many for sale to collectors in the international trade in rare species. And the angonoka's habitat was being overrun with bushpigs, imported from Africa. The local people believe that keeping an angonoka with their chickens will sustain the birds' health—strangely, people in south Madagascar keep a closely related species, the "radiated tortoises," with their poultry for the same reason. Maybe there is some truth in it.

In 1986, the Durrell Wildlife Conservation Trust (DWCT) launched Project Angonoka in collaboration with the Malagasy government and

Don Reid, whose name will forever be associated with the restoration of the angonoka, shown here with a female in northwest Madagascar. A radio transmitter is glued to her shell. (Don Reid)

with support from the World Wildlife Fund (WWF). For more than ten years, this program was headed by Don Reid, whose name will forever be associated with the restoration of the angonoka. I spoke to Don over the telephone, and he told me that when he first arrived at the site he found himself in a small field station, in the middle of a forest, surrounded by villagers who were not only puzzled by what these conservationists were up to but also suspicious of almost everything done by white people. There were a few WWF biologists who occasionally came and went, and although they were near a main road, it was extremely difficult to travel during rainy season. His job—to start a captive breeding program to try to save the angonoka from extinction.

Trial and Error

"When we started," Don told me, "we had to do everything from scratch. No one knew anything about the behavior of the tortoises. We didn't know what their diet was. So we had to go out plant collecting into the forest and sort of guess what they might like." They learned through trial and error. They found that the tortoises loved an introduced cactus. "They loved it so much we could give them medicine with the leaves," said Don, laughing. He told me that he found them strange creatures. "They sat around, doing nothing, for weeks throughout the long dry season," he said.

They started the captive breeding with eight individuals, five of whom were males, all of whom were confiscated from local homeowners. Gradually, over the years, others were confiscated, babies hatched, and the captive population grew.

Between January and July, each female digs between one and seven six-inch-deep nests, twenty-eight to thirty days apart. "She nests only at night," Don told me. "In each nest she lays only one, huge, egg. At midnight." Amazingly, all the eggs hatch within two weeks of one another, in the height of the wet season.

In November 1987, the first year of the breeding program, when Don went out at lunchtime to check the temperature (as he always did three times a day), he noticed that the soil in the center of a nest hole had sort of collapsed. "I saw a movement," he said. He fetched a spoon and very carefully felt under the sand—"and out came a baby hatchling!" To be followed by many more.

The First Step Is Trust

The other person with whom I talked at length was Joanna Durbin, who became involved with the angonoka program in 1990. She told me about the fascinating experiences she and others of the team had as they struggled to gain the trust, the interest, and finally the support of the local villagers.

At first, she told me, their only interest in the tortoises was to keep their chickens healthy. They certainly were not interested in conservation. Joanna was told that she should ask advice from the village elders, who told her (once they agreed to talk to her at all) that it was necessary for the conservation team to be accepted by the ancestors. She learned that King Ndranokosa, the last king of the region in the nineteenth century, often returned to his people and spoke to them through the voice of an elder. Quite often he attended village ceremonies.

One day, Don took her to a village where a sick person was seeking help. For a day and a night, they sat and watched. There was much chanting, some villagers went into a trance-like state, various people from the past appeared, and old women became young men. After that marathon introduction, it was not long before Joanna met the king himself—speaking through an elder, of course. It was a successful meeting, at the end of which he pronounced that the conservation team should be accepted since they were friends of angonoka. A cultural event should take place to bring the villages together to discuss the need for conservation of the angonoka and its habitat. They should hold a festival.

Eventually all was ready. The space was cleared in the traditional fashion by driving a huge herd of cows, some from each village, around and around through the undergrowth. There was singing, dancing, chanting, and a huge feast attended by the king himself. Remembering the trouble I had when I included, in my Gombe budget, money for sacrificial chickens, white robes, and so forth in order to exorcise black magic from our field site in the north, I asked Joanna who footed the bill. "The village elders organized it—the Durrell Trust paid for it," she said.

When, eventually, the time came to discuss the angonoka, it was decided that an area in the very heart of its habitat should be protected. "We used to manage our environment," said one of the elders. "We know how to do it. But no one bothers anymore."

The habitat of the angonoka is in a remote area 150 miles from the breeding center. It was, Don told me, just too remote to site the center there. He did some fieldwork there himself, but the detailed study of the angonoka in the wild was carried out by Lora Smith. Her work—also part of the Durrell program—on the behavior and requirements of the tortoises enabled the team to locate the best area for the establishment of a protected habitat.

A Trance, a Prayer, and Then a Release

Of course, the ultimate goal of the breeding program—to put angonoka back into the wild—could not happen until there was enough safe habitat to make this feasible. So it was a good day when, in 1998, the Baly Bay area in northwest Madagascar, optimum habitat for the tortoises, was declared a national park. It would be protected by eight full-time guards and a network of forty village para-rangers watching for poachers and forest fires, working closely with the local police.

Initially, a small number of juveniles were released and monitored. They adapted immediately, and their growth rate equaled that of their age mates in the breeding program. There were no deaths, no poaching, and no serious fires.

The first large-scale release took place at the end of 2005, when twenty young angonoka were released into large temporary enclosures in the forest. The event is described in a newsletter of the British Chelonia Group (BCG), an organization dedicated to promoting the interests of tortoises and turtles that raises money for conservation projects worldwide.

"We got to the village at dusk to a tumultuous welcome from the villagers who led us to a special palm thatched shelter decked out in greenery and flower chains," wrote Richard Lewis, the conservation coordinator for the Madagascar Programme of the Durrell Wildlife Conservation Trust. After speeches and an all-night dance (for those up to it!), the team and the tortoises finally set off for the forest the next morning. Everyone gathered at a small field station built at the edge of the forest. A spiritual leader offered a prayer, asking for the goodwill of the king and the ancestors. One of the elders went into a trance and spoke as the king, accepting the efforts being made by the conservation team.

Finally the twenty young tortoises, oblivious of all the hard work, planning, and celebrations, were taken into the forest and put in groups of five in outdoor enclosures. There they stayed for a month, getting familiar with the new habitat before being released, equipped with radio transmitters stuck to their shells with glue.

Over the next few years, more angonoka will be released into the wild. The success of the program is, of course, due to the commitment and dedication and hard work of so many people, including and especially the Durrell team—and Don Reid. And it is a program that will not be sustainable without the continuing goodwill of the local people.

MY CHILDHOOD TORTOISES

Writing this story brought back memories of my own two tortoises (not ploughshare!), which I had as a child. We had no knowledge of the pet trade that was endangering them in the wild, or the terrible conditions of their transport. The male, Percy Bysshe (because, with schoolgirl humor, he was "shell-y!"), was the first to arrive.

One day, despite searching everywhere, it seemed he had escaped for good. To our amazement he turned up about six weeks later—followed by a female! How on earth he had found her, I cannot imagine, since tortoises were not very common in our area. I named her Harriett, and they became an all-but inseparable pair. I suppose when she was receptive, he would follow her closely; when he got close behind, he withdrew his head and lunged forward to bump her shell with a loud *crack*.

It seemed that he always became particularly amorous when my grandmother was entertaining in the garden at teatime. Then, when she failed to divert their attention, the little group of ladies, despite their Victorian sensibilities, would be riveted as, again and again, Percy struggled to mount the impregnable wall of his beloved's shell, only to fall back as she, fed up with the whole procedure, simply walked away from under him. It's a hard life, being a tortoise!

My son rescued two females, with damaged shells, from the last shipment imported into England. When one died, the other seemed listless, and we thought she might die, too. To our amazement, she was befriended by the small black cat from next door. Day after day we would see him, curled up beside the lonely tortoise in her hutch. Eventually she went off to a colony in Chester Zoo, where she has adapted well.

Formosan Landlocked Salmon
(Oncorhynchus masou formosanus)

I heard about this fish in 1996 during my first visit to Taiwan. I had gone there at the invitation of Jason Hu, at the time the director of the Government Information Office, responsible for foreign affairs. A father of two children, caring passionately for the environment, he felt that a high-profile visit of someone well known to the international conservation community would help him in his efforts to better protect the environment. I was able to have meaningful talks with key decision makers, there was a good deal of positive media coverage, and finally, just before I left the country, I was given an audience with Taiwan's President Lee Teng-hui.

It was a positive meeting. We talked about animals, the environment, and various conservation issues. I showed him some of the symbols I carry around the world—such as the flight feather of the California condor—and I asked him if he could think of anything I could take away to symbolize a Taiwanese success story. This is when he told me about the fight to save the Formosan landlocked salmon from extinction. I was fascinated by the story, but felt it would not be quite appropriate to travel around with a dried fish, as he suggested!

Survivor from the Ice Age

The Formosan salmon became landlocked during the last ice age, trapped in cold mountain streams. It is found only at elevations above five thousand feet in the Chichiawan Stream, where water temperatures can fall below sixty-five degrees Fahrenheit. Studies show that the salmon

The joy of saving a unique species. Dr. Liao Lin-yan, a leading force behind Taiwan's effort to save and restore the Formosan landlocked salmon. *(Liao Lin-yan)*

requires this exact temperature of extremely clean-flowing stream to survive.

At one time, it was plentiful and was a staple in the diet of the aboriginal people living there, who called it the bunban. But by the end of the last century, due to overfishing and pollution, there were only some four hundred individuals known to exist, making it one of the rarest fish in the world. If it was not to slide into extinction, something would have to be done.

And in the late 1990s, something *was* done. A dedicated team from the Shei-pa National Park decided to protect and restore the fish. An important member of this team, Liao Lin-yan (a PhD candidate at the time), is especially committed to the cause. Unfortunately, during my last visit to Taiwan I was not able to meet with Dr. Liao Lin-yan. And as he cannot speak English nor I Chinese, we could not even talk on the telephone. But Kelly Kok, executive director of JGI-Taiwan, talked with him and translated the information he offered.

Liao has loved animals since he was young. He initially wanted to be a veterinarian but was accepted into the department of aquaculture. "It's quite the same thing really," he said. "Fish get sick, too, and instead of helping individual animals, I get to help a whole pond of them!" After graduating, he applied successfully for a job at the Shei-pa National Park.

"Because the Formosan landlocked salmon is a finicky species, needing clean water at the right temperature, and very particular about its diet, trying to restore them to the natural environment is difficult," said Liao, "but our team was determined to make it happen." They racked their brains to think of ways to increase the number of fry, and then spent sleepless nights persuading the young fish to feed. "The trouble is, they prefer live organisms," Liao explained, "but water fleas are difficult to acquire in the mountains and shrimp is not an appropriate choice." And so they had to be trained to eat fish food. Liao described how he made the food float about on the water, looking like live prey. "And once the more daring ones take their bites, the timid ones follow suit," he said.

When Liao began working on the project, the condition of the restoration ponds as a result of frequent typhoons and flooding was appalling—they looked like abandoned pits. And the equipment was inadequate and makeshift. "We really had a hard time back then," Liao recalled. "The ponds were located far up in the mountains, and acquiring the most simple maintenance parts was a task." Nevertheless, gradually they improved conditions.

Risking Their Lives to Save the Salmon

And then, in 2004, exceptionally strong typhoons struck Taiwan—and the team, out in the field, had to watch helplessly as the level of the water rose in the ponds, and precious fish were carried away by the floods. They battled to save as many as they could, but it was dangerous work and they were risking their lives—at any moment one of the team could have been swept away by racing floodwater. And not only were they losing fish, but both water and power supplies were cut off. The remaining fish were endangered by leaking tanks and by rising water temperatures. The team brought in an emergency water supply truck. They borrowed ice blocks from local hotels. The hard work of the previous few years had mostly been swept away. "It was a pity," Liao said, "but we can always start all over again."

In fact, because of the disaster, the restoration team realized that the precious fish needed a safer environment, and they decided to establish the Formosan Landlocked Salmon Ecological Center in the Shei-pa National Park. It was not easy to raise the money, but after months of hard work they succeeded. In 2007 the center, with its comprehensive facilities to ensure constant power and water supplies, was complete. It was launched with a big celebration at which a dance troupe from the local elementary school performed traditional dances and sang tribal songs of the Atayal aboriginals. Each guest was given a sapling to plant to symbolize land rehabilitation and the protection of the bunban. "With teamwork, we can help the salmon fry breed in the ecological center and maintain the population at five thousand," said the director of the park.

That celebration launched the first of the center's educational programs, Dialogue with Mother Nature, emphasizing the importance of protecting the environment and protecting endangered species—especially, of course, the "National Treasure Fish," the Formosan landlocked salmon. Over the years since the project began, many scientists have been involved from national universities; the little fish has created a flurry of interest spanning many aspects of its prehistory, ecology, and behavior. The Taiwanese are proud of their unique landlocked salmon and are determined to do their best to protect it.

The Need for Backup Populations

Clearly there was a danger that the one population of Formosan landlocked salmon could be wiped out by typhoons or disease, and the recovery team decided it was necessary to try to establish additional populations as backups. After extensive surveys, they found two areas that seemed suitable, in the Szechiehlan and Nanhu Rivers. Dr. Liao told me that some one thousand fish have been released into the new sites. A survey conducted after two years found schools of eighty to ninety salmon in the Szechiehlan River containing second-generation fish; in the Nanhu River, about forty young fish were found.

Thus the outlook for the long-term survival of the Formosan landlocked salmon gradually improved. By 2008, the national parks team had worked for ten years to protect the Chichiawan River Basin area, and the number of salmon there has remained fairly stable, at around two thousand individuals, over the past few years. A big difference from the few hundred that existed at the start of the restoration project.

A Very Special Memory

I asked Kelly to contact Liao one more time to see if he could share with us a personal story. He wrote about something that happened when he joined the team in 1999. "The first time I attempted to capture a salmon for artificial insemination, I was very concerned as to whether I would get the right fish—a female that we could successfully inseminate. There were about five hundred wild salmon out in the Qijiawan Creek—how could I be sure that I would get the right one? I faced my task with trepidation and hoped for all the luck I could get."

By then, the team had discovered that the best way to capture a mature salmon was to use fishnets, working at night so that no shadows were cast on the water. On this occasion, Liao was in charge of the group that headed for the creek to cast the nets; another group was waiting in the lab to receive any fish caught. "We cast around for three full hours on that trip," Liao remembered. "Every time we cast the net, we checked our results eagerly." But again and again the nets were empty, and at the end of the three hours they had caught only one salmon, a male. As they were all absolutely exhausted, they decided to call it a night.

"But then," said Liao, "I spotted a female salmon lying near the bank; she was apparently full of eggs. I caught her, and together with the male fish, we rushed back to the lab." The excitement and suspense came through along with the words. "It turned out that my salmon was ready to lay her eggs. All we had to do was gently squeeze them out and inseminate them. There were over six hundred eggs! I was told that it was the most beautiful female salmon they had ever seen. It was truly my lucky fish!"

Because of his work, Liao has had to spend a lot of time away from his home in Keelung, which is a three-hour drive away. But he said that, although his work is hard, "the sight of a freed salmon swimming with ease in the river more than compensates me for my sacrifices." Liao is always reminding everyone that each one of us must do our bit to help protect our planet. "Fish cannot survive in polluted waters," he told me, "but neither can we!" And he speculated that "if the Formosan landlocked salmon becomes extinct, it may be that human beings will eventually disappear from earth as well."

Vancouver Island Marmot
(*Marmota vancouverensis*)

These marmots, about the size of a domestic cat and weighing between five and fifteen pounds, are extraordinarily attractive. With their thick chocolate-brown fur, white muzzles, and engaging expressions, they look just like characters in a Walt Disney classic. Historically, they lived on Vancouver Island in the sub-alpine meadows that are created and maintained by avalanches, snow-creep, storms, and fire. Such meadows tend to be rare on Vancouver Island, which is why these marmots have never become more plentiful.

It was not until 1910, when some were killed to provide museum specimens, that the Vancouver Island marmot was recognized by science as a species. After this, confirmed sightings were rare until 1973, when Doug Heard of the University of British Columbia began studying marmot behavior in two colonies. Several years later, local naturalists began systematic population counts. And in 1987, Andrew Bryant began a study that, originally planned as a short-term master's project, has already lasted more than twenty years! It has involved him, deeply, in efforts to prevent the Vancouver marmot from toppling over the brink into oblivion.

Meet Andrew Bryant, Marmot Man

In December 2007 I called Andrew, who lives on a small farm near the city of Nanaimo on Vancouver Island, and we talked—about marmots—for a long time. As with so many of the other individuals featured in this book, I longed to sit down with him and have a face-to-face conversation. And in November 2008, I was able to do just that over a quiet lunch at

Andrew Bryant has devoted his life work to protecting these enchanting mammals. Shown here with Barbara at Pat Lake, Vancouver Island, British Columbia. *(Andrew Bryant)*

the hotel where I was staying in Vancouver. Originally I had planned to visit the marmots themselves—but it was the wrong time of year, and they were all hibernating. In fact, they hibernate for many months each year. So he brought me a gift of a stuffed toy marmot instead! And in fact, having the chance to sit and talk with the man who has worked so hard to save them was, in a way, even better than seeing the marmots. For he is a man after my own heart, with a great sense of humor, an obvious passion for his work, and a deep love for the marmots to which he has devoted so much of his life.

Clear-Cutting on Vancouver Island

In the spring, Andrew told me, young marmots, during their "teenage years" (usually when they are two years old), traditionally left their mountaintop homes and traveled to the tops of nearby mountains where they bred and lived for the rest of their lives. But by the time he began his study, years of unregulated clear-cutting of the forests during the 1970 and 1980s had led to a change in marmot behavior. The timber companies, inadvertently, had produced new habitats similar to the marmots' preferred natural one—treeless regions that are open, covered with grasses and other plants to eat, and with good soil in which the marmots can make their dens.

And so many of the adolescents no longer bothered to journey to the next mountaintop, instead colonizing the clear-cut areas. The additional habitat enabled their population to boom, likely doubling from 150 to 300–350 by the end of the decade. "If it didn't influence predator populations, clear-cut logging would likely be beneficial for marmots," Andrew told me.

However, the marmots preferred to use logging roads to move about in their new habitat, which made them easy prey for the many cougars, wolves, and golden eagles that also lived there, especially where the trees had started to grow back, thus providing cover. The rich new areas, Andrew told me, "became 'population sink,' since few of the marmots who moved there survived for long." And so the wild population plummeted. By 1998, there were only about seventy marmots in the wild, and five years later the number had dropped to thirty. Ironically, a contributor to this final decline was the collection of marmots for captive breeding programs—fifty-six wild-born marmots were taken into captivity between 1997 and 2004.

"If It Hadn't Been for the Logging Companies . . ."

In the early days of his study, Andrew was heading up the mountains on a nearly daily basis to observe the marmots on the privately owned logging land they inhabited. He would often sign in, at around 3:30 or 4 AM, as "Mr. Marmot, heading up Green Mountain." As the months went by, the loggers became curious as to what he was doing on their mountain in the early hours, and one morning Wayne O'Keefe, a logger who worked for MacMillan Bloedel Limited, decided to drive up and see what "Mr. Marmot" was up to.

"All of the elements were perfect that morning," said Andrew. "I was able to tranquilize a female and tag her, and get a photo of Wayne holding the marmot with his wonderful Day-Glo safety vest and hard hat

Wayne O'Keefe, a logger who worked for MacMillan Bloedel Limited, holding Iris. This was the moment that led to major changes in logging practice to benefit the Vancouver Island marmot. *(Andrew Bryant)*

glistening in the light. The man said, 'Wow, this is cool—you should come down and talk to the guys over lunch.' So I did."

The talk was a big success, and led to his meeting with first the logging foreman, then the woods manager, and finally with Stan Coleman, a highly placed manager within the company. "I found myself in the boardroom of a logging company, armed with photos and slides and maps, and told the man what I thought I knew about marmots and the effects of logging." After listening patiently, Stan asked, "What do you want me to do about it?"

"I told him he could take ethical responsibility for the animals now, or have their demise on his hands when the species is lost," Andrew said. After that meeting, Stan became one of the marmots' greatest advocates, encouraging the company to do all it could to support marmot conservation efforts.

Andrew loves to point out that, while the Vancouver Island marmot was the first species to be officially listed as endangered in Canada, and was also listed as endangered by the IUCN and by the US Fish and Wildlife Service, these listings "ultimately did little or nothing to help marmots—but the love-of-the-landscape by a local woodcutter eventually led to something that did." A donation of a million dollars Canadian helped establish the Marmot Recovery Foundation (MRF), a nonprofit organization to which Andrew continues to serve as scientific adviser.

The lands were later purchased by Weyerhaeuser in 1999 and subsequently resold. Today the marmot habitat is owned by Island Timberlands and Timber West Forests. So far all of the landowners have honored the commitment to marmot protection in the form of ongoing financial support. "And the great irony of it all," Andrew says, "is that if it hadn't been for the logging companies, we never would have gotten started!" He told me that, while logging still continues, there are protected areas now; as the forests regenerate, conservation measures can be implemented.

The Recovery Work Began in Earnest

The Marmot Recovery Team estimated that a total population of four to six hundred marmots would be necessary, dispersed in three separate locations on Vancouver Island, to constitute full "recovery" of the species. From 1997 through 2001, the team drew up a recovery and reintroduction plan that included a captive breeding program to be implemented by the Toronto Zoo, Calgary Zoo, the privately owned Mountain View Breeding and Conservation Centre in Langley (near Vancouver), and a specially designed facility constructed at Mount Washington on Vancouver Island. Release sites were selected where the habitat was appropriate and predation less prevalent. The first pups were born in the breeding program in 2000, and the first of these were released into the wild in 2003.

Field observations after release showed that captive-born marmots had retained natural predator-recognition skills, often whistling to alert other marmots when a cougar or wolf approached, or when an eagle flew overhead. In 2004, the team achieved something of a milestone when a captive-born marmot mated with a wild female, and the following year the first litter from a captive-born pair was recorded. They were living in a fully protected natural habitat where an earlier colony had become extinct.

When I met Andrew, he told me that 2008 had been a hugely successful year for the marmots. The captive breeding program had "the largest litter ever recorded—nine pups weaned—and fifty-seven marmots were released into the wild." And eleven litters totaling thirty-three pups were recorded in the wild, most of them from captive born parents. As of October, there were 190 individuals in captivity and approximately 130 in the wild. "The grand total of 320 marmots is a far cry from the estimated total of 70 individuals in 1998," Andrew said triumphantly. Moreover, these individuals are distributed over a dozen mountains—whereas five years before, no more than five mountains had been occupied.

Each wild pup is tagged, each individual of the population is known—all of them by number, many still by name. And the genetic profile of each individual is known, too. The Marmot Recovery Team is justly proud of the fact that genetic variation has been maintained—that not one allele has been lost since the start of the captive breeding program (although some genetic variability was undoubtedly lost when whole colonies became extinct).

The recovery plan requires captive marmots to be released each year to boost the wild populations until a self-sustaining population of four to six hundred marmots is achieved—which should be sometime during the next fifteen to twenty years, Andrew believes, although he points out that many factors have to be considered in a long-term perspective—such as numbers of predators, continuing support, and global warming. Nevertheless, all in all he remains highly optimistic that the Vancouver

Island marmot will be fully restored to its traditional habitats on the mountaintops.

Oprah Winfrey, Franklin, and All the Rest

And of course there are the marmots themselves, each with his or her own personality and contribution to make. When Andrew began writing up his data for his degree, his supervisors chastised him for using names instead of numbers to identify individuals. He told me that, for him, names were easier to remember—something I, of course, agree with! He knew the marmots as individuals: "I knew where they lived, I knew what they did, and I knew how to find them," he told me. He named his favorite female Oprah Winfrey, and he knew her for ten years, during which time "she had eleven pups before she was killed, probably by a wolf." Then there was Franklin, who was tagged as a (nameless) pup, monitored for a while, but by the following year had vanished. Five years later he turned up, alive and well, on Mount Franklin—hence his name. "Since then," Andrew told me, "Franklin has sired a whole gaggle of pups."

Andrew is unstinting in his praise of all who have helped make this recovery effort possible. "I had the luxury of some rather tall shoulders to stand on and a large number of supporters. More importantly, without the dedication of a huge and very talented array of people—staff and volunteers—dreams would have remained just that. I can't stress this enough . . . it is *teamwork* that has given this species a potential future!"

But no one has done more to ensure the survival of these delightful creatures than Andrew himself. I asked what kept him going during the past twenty years, through the downtimes when things go wrong He smiled as he answered, simply: "I truly love those little guys. They are real survivors: They have learned to live where few creatures dare."

THANE'S FIELD NOTES

Sumatran Rhino
(*Dicerorhinus sumatrensi*)

It was only hours after his birth that I first laid my eyes on Andalas, the baby Sumatran rhinoceros who was born in fall 2001 at the Cincinnati Zoo. After years of anticipating this miraculous birth, I was amazed by how drop-dead cute he was, with his over-size eyes and surprisingly thick red hair. Rhinos are many things, but "cute" is not a typical field marking for the animal. This was the first successful birth of a captive Sumatran rhino in 112 years. Nicknamed the "hairy rhino" for its long red hair, the Sumatran rhino is the most endangered large mammal in captivity. The species is down to fewer than three hundred in the endangered wilds of Malaysia and Indonesia.

This is an animal that has hidden its secrets deep in shadowed forests for a very long time. It has taken a lot of big brains, and even heftier hearts, to understand the ways of this elusive species in the hope of keeping it in our world.

In fact, six years later when the world's third Sumatran rhino was born, the *Today* show sponsored a naming contest. The winner was *Harapan*, which is an Indonesian word for "hope." I can't think of a more fitting name, as "Harry" is indeed a symbol of hope for this beleaguered species. He won all our hearts at the zoo as soon as he began tentatively walking just an hour after his birth. And we had no doubt he would grow to be a big boy, since he nursed every fifteen to thirty minutes for his first few weeks of life. (The growth rate of this seldom documented species is astounding: Harry weighed eighty-six pounds at birth, and topped two hundred at four weeks of age. When he's a full-grown adult, he'll weigh about fifteen hundred pounds.)

The saga of how Andalas, Harry, and their sister, Suci, finally came to be born is an excellent example of how challenging captive breeding can be. In 1990, a bold plan to help save the Sumatran rhino was launched by a consortium of American zoos working together with the Indonesian government to form the Sumatran Rhino Trust. The plan was to import rhinos for captive breeding from the areas of forest in South Asia that were slated to be cut down for

timber and cropland. Just seven rhinos were shipped to leading
American zoos in an attempt to not only breed the species as insurance against extinction, but also raise the public's consciousness to
the plight of the wildlife of South Asia.

From the get-go, the captive breeding plan was controversial.
Much like when the California condor was taken out of the wild
for breeding, there were some conservationists, including those
from the Asia-Pacific branch of the World Wildlife Fund, who
fought the captures. One of their main concerns was the near-total
lack of understanding of the husbandry requirements for this
elusive species.

As it turned out, this lack of husbandry knowledge contributed to
the deaths of several captive rhinos, which were originally imported
in 1990. Figuring out how to care for these surprisingly delicate
beasts was much harder than anyone planned. When the rhinos first
came to America, everything seemed to be going splendidly. They
were eating enormous amounts of timothy hay and alfalfa, and
though they are antisocial by nature, it was hoped they would settle
in and breed well in zoos like their African rhino cousins.

However, the Sumatran rhino was not a plains rhino, accustomed
to digesting grasses. They come from the secluded reaches of
dense tropical rain forests, and their habits and food preferences
were little known. So even though they were eating, they were not
getting everything they needed from their hay and grain. Soon
many of the rhinos brought into zoos were ailing. And by four years
later, in 1994, there were only three hairy rhinos left in captivity, all
at the Cincinnati Zoo.

Then it came down to a life-altering moment of truth. At that
time, in sad and sometimes contentious meetings, zoo veterinarians, keepers, and Zoo Director Ed Maruska discussed what they
would do for Ipuh, who had not eaten or stood up in days. He was
literally withering away. After much arguing and wrestling with the
issues, it was decided that this species was too rare and Ipuh too
valuable as a potential breeder to consider euthanasia. However,
something had to be done.

And the solution was surprising. The head keeper of the rhinos—
a gruff, tattooed rhino of a man named Steve Romo—though not a
nutritionist, was the one to solve the puzzle of Ipuh's health. Romo,
as he's called at the zoo, put it this way: "I knew from watching

Mahatu, our female, waste away on her diet of hay and pellets and die that this did not work for Sumatran rhinos."

Romo told me he learned about the Sumatran rhinos' natural diet in 1984, when the US Zoos sent him to Malaysia to assist with the first Sumatran rhinos they rescued, Jeram and Eronghe. He remembered that Jeram ate food with "a lot of very sticky sap in the jackfruit . . . the same sticky sap that is in ficus." Although jackfruit could not be found in the United States, Romo knew ficus could.

"No one expected Ipuh to survive, including myself," Romo told me. So he ordered some ficus for Ipuh's "last meal." However, when Romo dragged the ficus into the barn and started washing it off, the keeper sitting watch with Ipuh yelled, "Hey, I don't know what you've got, but Ipuh lifted his head for the first time in two days!"

From forty feet away and around a stall door solid enough to contain a male rhino, Ipuh could smell the ficus. And when they brought it to him, Ipuh stood up and began to eat. In fact, he ate it all in just two days and continues to this day to eat fig and ficus species flown in from California in refrigerated boxes. Which also makes the Sumatran rhino by far the world's most expensive animal to feed!

Ipuh, now thirteen years later, remains the only captive breeding male Sumatran rhino in history. He has sired three young, including our beloved Harry, and continues to thrive. And what Romo taught us about rhino nutrition has been put into place across the globe, from the zoos that exhibit Sumatran rhinos to a protected area in Indonesia, where small captive populations are kept on the edge of the preserves.

And to this day a partnership has been underway with the San Diego Zoo, which collects ficus and fig brows and ships them to the Cincinnati Zoo to feed the Sumatran rhinos.

BREEDING MISHAPS AND MYSTERIES

Figuring out how to feed Ipuh was the first mystery to be solved. Figuring out how to successfully breed him with a female was almost as deadly. Every single time keepers put the male and female in the same yard, they fought and chased and screamed and smashed into each other, often until they were bleeding. You can imagine the mayhem outside the rhino yard, as well. I can assure you that the rhinos weren't the only ones screaming.

So year after year, attempt after attempt, their keepers would intervene and have to use fire hoses to separate them. And this went on for five years, until the Cincinnati Zoo hired a young reproductive physiologist named Terri Roth, who had been working at the Center for Research and Conservation at the National Zoo in Washington, DC. Months passed as Terri and her team studied the hormone levels in Emi's urine and feces. Eventually the veterinary technicians conditioned Emi to let them draw her blood and perform daily ultrasound examinations of her ovaries. At last, in 1997, Terri's crew determined that Emi's estrus, or receptive period, was only about twenty-four to thirty-six hours long, so pinpointing it was essential for successful mating.

FIRE HOSES READY

Terri was able to identify the exact day that the keepers should put Emi and Ipuh together for mating. That morning everybody was nervous and excited. And, of course, keepers were stationed with fire hoses on the sides of the yard, as always. But as of that spring morning the fire hoses have been relegated back to the hydrants, because those rhinos put on a surprisingly amicable demonstration of reproduction.

Ipuh tried to breed Emi forty-seven times that day and was never quite successful, but there wasn't a single chase or fight to worry about at all. Twenty-one days later, Terri had the rhinos put together for breeding again, and this time Ipuh was successful. Soon it was determined, to everyone's delight, that Emi was pregnant. However, the challenges were not yet over. Over the next few years, Emi developed a pattern of becoming pregnant, then miscarrying within the first ninety days. She eventually lost five pregnancies between 1997 and 2000, prompting Terri to prescribe a daily oral dose of progesterone, a hormone she knew was commonly used on horses, for Emi's next pregnancy.

And it worked! On September 13, 2001, a male calf, named Andalas (one of the original names for the island of Sumatra), was born at the Cincinnati Zoo. Emi, it ends up, after all these trials and tribulations, is a phenomenal mother. Andalas weighed seventy-two pounds at birth and stood up and walked at fifteen minutes old. He nursed like wild and reached nine hundred pounds by his first birthday. After four years at the Los Angeles Zoo, Andalas was sent to Indonesia, where a small captive population of hairy rhinos is kept on the edge of a preserve.

The return of Andalas to his native land not only met with international media attention but also brought the story full circle, demonstrating that captive breeding of this critically endangered species is not only possible but looks to be succeeding. The hope is to establish captive breeding populations adjacent to the parks so that the young can be more easily released in the wild to bolster the population.

Meanwhile, in the past few years Emi and Ipuh have continued to breed successfully. Now that Emi's a veteran mom, she no longer needs progesterone to carry her fetus to full term. The immediate plan is to breed Andalas with two young females at the sanctuary in Indonesia to increase the genetic diversity in captivity.

So does breeding a few rhinos save a species? Not by itself. Demonstrating the value of wild animals to the people in an area where they live is the number one way to help protect those species and their habitat. Perhaps the most important outcome of this successful captive breeding program is the increased public awareness and dedication to protecting the wild hairy rhinos—around the world, but especially in Indonesia.

THANE'S FIELD NOTES

Gray Wolf
(Canis lupus)

The first time I saw wild wolves was in the Lamar Valley in the northern section of Yellowstone National Park. I am willing to say that I believe in miracles solely on the basis of seeing those wolves. Because it is nothing short of a miracle that the gray wolf—once ruthlessly eradicated from the region—has returned to Yellowstone.

Long feared as a vicious predator, this social creature is actually very important to people. Not only are wolves the ancestors of our beloved dogs, but they are a remarkable symbol for conservation. In the Greater Yellowstone Ecosystem in the American West, the successful reintroduction of the gray wolf has been both a remarkable comeback and a major controversy. Perhaps the greatest miracle is that ranchers are starting to live together with wolves for the first time since the advent of guns. And even though some ranchers continue to resist the reintroduction, the wolves are back, and by all appearances they are here to stay.

As a result, the return of the gray wolf to Yellowstone National Park is number one on the official "World's Top Ten Conservation Programs" list. It took decades of work and educating and arguing and explaining for that crazy combination of Western conservationists, ranchers, federal biologists, and dreamers to succeed.

In my mind, what makes wolf restoration so significant is that it follows hundreds and hundreds of years of persecution. There are documents from the 1600s offering bounties on gray wolves in a number of colonies, stating the desired eradication of the species from the land. For centuries, Americans worked diligently to kill the wolf. And by the early twentieth century, the job was done. A species that once lived throughout most of the lower forty-eight states was almost entirely gone (a small remnant population was able to survive in Minnesota).

This almost complete eradication was not due just to the use of leg-hold traps. Or to wanton hunting and bounties. What finally took out the wolves was the widespread use of poison over a huge landscape. What brought them back was a public outcry and backing of efforts to restore the wolf to suitable habitat in the American West.

Mike Phillips is the executive director of the Turner Endangered Species Fund, headquartered in Bozeman, Montana. Mike ran the gray wolf restoration program in Yellowstone National Park from 1994 to 1997. But a decade before that began, he was involved with the reestablishment of the red wolf to the American Southeast, which Jane speaks to in these pages. Mike told me that during his many restoration projects, he learned a great deal about communicating with, and especially listening to, the concerns of local people.

"You have to make sure that local folks know exactly what you're up to because they will ask you, 'No, you want to do *what*?!' And ranchers throughout the West will often say to me, 'It's not so much that we're against the gray wolf. We can live with the wolves. But what we do not want is a further erosion of our way of life. We don't want further federal intervention into what we do. We don't want further state intervention into what we do.' They see this as just another indication of how the West they once knew is changing."

Of course, the effort isn't to restore wolves to private ranches, but on federal land in Yellowstone National Park, where national surveys had shown that an overwhelming majority of Americans wanted again to see wolves thrive. What heartens Mike Phillips is "the continued bipartisan support of the Endangered Species Act for more than thirty years. The law is much argued over and debated, but the American people steadily tell their representatives that they do not want and will not accept extinction on their watch in their country."

Mike was there at the cages on that March morning in 1995 when the first wolves in sixty-nine years were allowed to run free in Yellowstone. What made the project so complicated was that wolf restoration wasn't just a sociopolitical challenge, or an administrative one with all the logistics, but also a biological challenge.

"We knew from other wolf relocations that if we simply released the gray wolves in Yellowstone, they would take off," Mike told me. "But the purpose of this program was to restore wolves to Yellowstone, so we needed to release them in a way that they would have a strong tendency to stay in the region. So we had to arrange an acclimation program that would allow the wolves to remain in captivity for an extended period of time at the release site. Well, that meant that they had to be fed and watered."

Water was the easy part of course. "For heaven's sake, in wintertime they can simply eat snow," Mike told me. "But imagine the amount of food we had to provide. We needed to feed five pounds of food per wolf per day. And if you have twenty wolves to be released, that's a hundred pounds a day and seven hundred pounds a week and three thousand pounds per month. It starts to add up."

Mike has been surprised by the success of the wolf restoration. "On any measurement you would like to observe, the program has been a success. The population has grown faster than expected. And particularly surprising is that many of the packs have remained observable. It is rather routine today for visitors to see wolves in the wild in the Northern Range of Yellowstone Park."

And as for what the future of wolves looks like, Mike reveals that at his core, he is a biologist: "The other thing you have to be mindful of is that gray wolves are great ecological generalists. They don't need much of an opportunity to flourish. They largely need to be left alone and they need access to prey items that are typically bigger than themselves. You give wolves a big landscape with something to eat, they're going to do just fine."

Basically, he is unconcerned about the wolf's future in Yellowstone and nearly unconcerned about the wolf's future in the northern Rocky Mountains generally. As a result of Mike Phillips's work and that of hundreds of others involved with the gray wolf restoration program, today you can see wild wolves again in the West, just as Lewis and Clark did two centuries ago.

PART 3

Never Giving Up

Introduction

So far our stories have been about species rescued from the very brink of extinction and reintroduced to nature, although very few of them are surviving with absolutely no human management. And with the prospect of continued human population growth, habitat loss, pollution, poaching, climate change, and so on, we must remain vigilant in our effort to protect them and their habitats.

Those grouped in this section have a future that is even less secure. They have been saved from toppling into the abyss of extinction but, for various reasons, they have not yet been reestablished in the wild.

In its vast desert habitat of Mongolia and China, the wild Bactrian camel is threatened by hunters—and also by lack of water, as so much of the snowmelt in the surrounding mountains is diverted for agriculture—and will, presumably, be diminished further by global warming. Its future will depend on continuing talks with the Chinese and Mongolian governments and the political will to find an area where the wild Bactrian camel will be safe and its needs met. The future of the Iberian lynx in the wild depends on the extent to which the authorities are prepared to protect areas of natural habitat from human encroachment—and to some extent on the lynx's ability to learn how to cross roads safely!

Some must be retrained during captive breeding to adapt to the reality of their habitat. The giant pandas that are bred in captivity must be raised in such a way that they can survive and find more suitable food in their natural habitat than has been the case so far. And the effort to teach the northern bald ibis a new migratory route is still in the pilot phase—though this is very encouraging.

I have met many of those who are involved in the efforts to ensure a safer future for these species in the wild—some of them have been involved for many years. Fortunately for the animals—and for future generations—none of them will ever give up, no matter the challenges they face.

A further point to be made: These stories are representative of countless other rescue efforts that deserve to be publicized, some of which—such as the Chinese alligator—will appear on our Web site (janegoodallhopeforanimals.com). One of the problems I have faced, during the writing of this book, is just how many admirable efforts are being made to save endangered species, all over the world. Just today, for example, I read about the beautiful little ladybird spider that lives close to my home in the UK. Its numbers were once down to about fifty individuals, but thanks to captive breeding there are now a thousand. I hope, on our Web site, we can honor many more of these ongoing projects, and the scientists and citizens who are helping to maintain and restore the biodiversity of our planet.

We do not know what the future holds for life on earth, whether our combined efforts can turn things in favor of animals and their world. What is important is that we never give up trying.

Iberian Lynx
(Lynx pardinus)

I first read about the Iberian lynx in the Iberian Air magazine *Ronda Iberia* in June 2006, when I was on my way from Spain to the UK. Endemic to the Iberian Peninsula, this lynx, I read, is one of the world's most endangered felines. The article introduced Miguel Angel Simon, a biologist who was heading up a lynx recovery plan. Immediately I wanted to meet with him.

And a year later, when I was in Barcelona, it happened: Miguel Angel flew in from his field station to talk with me. I found him sitting at a table in a peaceful area of my small hotel with Ferran Guallar, executive director of JGI-Spain, who offered to translate for us. Miguel Angel, a wiry man with a short military mustache, looked business-like and competent, and was clearly passionate about his work with the lynx.

It was 2001 when Miguel and his team began the first thorough census of the lynx population throughout Andalusia. They set up photo traps and searched for signs of lynx presence such as feces. The results showed that the species was in serious trouble. Not only were the lynx affected by habitat loss, hunting, and being caught in traps set for other animals, but rabbits, their main prey species, had been almost eliminated by an epidemic. Indeed, in some places they had disappeared entirely from this land the Phoenicians called Hispania, meaning "land of the rabbits." Undoubtedly, Miguel said, many lynx had died from starvation. His census showed that there were only between one and two hundred lynx remaining in two areas in southern Spain; during the previous twenty years, they had become extinct in central Spain and Portugal. Clearly desperate measures would have to be taken if these beautiful animals were not to become extinct.

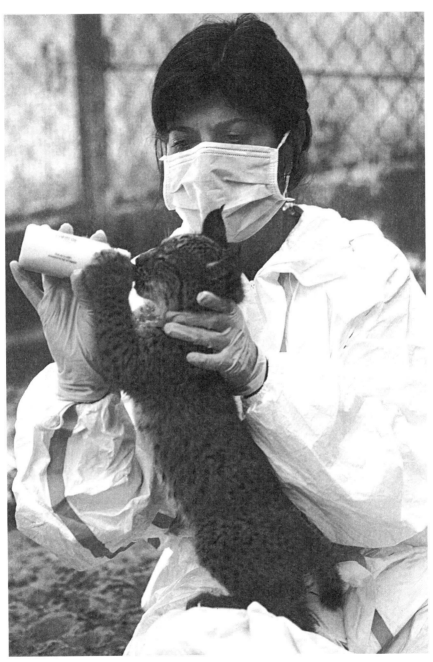

Astrid Vargas and her staff work around the clock to save Spain's treasured Iberian lynx. Shown here with a young male, Espliego, abandoned by his mother, Aliaga. *(Jose M. Pérez de Ayala)*

Winning Friends for the Lynx

An application to the EU for funding resulted in one of its biggest-ever grants for work with an endangered species—twenty-six million euros for the period from 2006 to 2011. The lynx restoration program was established with eleven partners: four conservation groups, four government ministries, and three hunting organizations. Because most of the surviving lynx were on private land in the rural boroughs of Andujar in Jaen, Cardena in Cordoba, and Doñana in Huelva, it was clearly of utmost importance to strive for the full cooperation of the landowners.

At first, this was not easy. A lynx will prey on fawns, and many farmers had concerns about lynx also killing their lambs—which they sometimes do. And so, from the start, Miguel and his team investigated every report of lamb killing and gave compensation to farmers—even if it turned out that the killer had been a wolf. A scheme was launched whereby awards were given to landowners who had good conservation records.

Gradually the landowners' attitude changed. More and more of them, whether they owned fifteen thousand acres, or fifty acres, or simply a summer villa with a garden, signed agreements with the lynx recovery team. First—they would protect the lynx on their land. Second—they would no longer shoot rabbits, but rather leave them for the lynx. Third— they would permit those working on the lynx recovery plan to use their land for controlled reintroductions (of lynx and rabbits) and monitoring. Indeed, it has become something of a status symbol to claim that you have lynx on your land—after all, in some places the lynx is actually a totem animal. Thus the lynx is now protected, through ninety-eight separate agreements, throughout an area of some 540 square miles.

Of course, Miguel told me, recovery is agonizingly slow. A female has cubs only every other year, and normally she will not raise more than two young at a time. Nevertheless, in 2005 at one of the main research sites, some twenty females gave birth to about forty cubs in the spring. And by autumn approximately thirty young lynx had survived. But this is the time, Miguel told me, when the trouble starts, as the young adults leave to find new territories. The males leave when they are a year old. The females may hang around for another season. Whatever their age, many simply vanish when they go off on their own. But according to Miguel, they have now started using radio collars for GPS satellite tracking; it is finally possible to find out where the animals go.

I asked Miguel if he had a good story to share and he told one that proves, he says, that the conservation program is working. In 1997, in one area, there were only seven adult lynx (identified from photo traps)—two females and five males—and just one cub. Nobody thought the tiny group had a chance of surviving, especially because disease was spreading among the rabbits. Nevertheless, the son of the ranger in charge was asked to name the cub. The little boy, without hesitation, chose the name *Pikachu*. And, to everyone's amazed delight, Pikachu—along with all seven adults—survived. Today there are forty-five lynx in the area. "And," said Miguel, "Pikachu is king."

A Visit with the Lynx

Written into the recovery program was the decision to establish a captive breeding program. A team of scientists, who work closely with Miguel and his team, carefully determine which lynx, from which areas, should be taken into captivity in order to ensure genetic diversity. The rules are strict: Only if three cubs from one female survive to six months of age can one of them be captured. The cubs are sent to one of the two breeding centers.

Miguel works closely with Astrid Vargas who heads up the El Acebuche Centre in Doñana—and he introduced me to her by phone. A year later I was landing in Seville, with my sister Judy, for the drive to the breeding center. Astrid herself was unable to meet us at the airport because there had been a tragedy during the night. She had been woken by the volunteers who monitor the breeding females and cubs via TV monitors. They told her that there had been a serious fight between cubs—the sixth within the past month. This time it was Esperanza's youngsters. By the time Astrid arrived, the female cub had received a lethal bite to her throat.

I learned that it was the second death caused by cub fighting since the start of the breeding program. Thus it was a somewhat subdued team that greeted us when we arrived: Astrid, Antonio Rivas (Toñe), Juana Bergara (the head keeper), and some dedicated volunteers. It was not surprising that they were upset—they showed me the footage of the aggression later, and it was shocking in its sudden onset and its ferocity.

Astrid told me that she could never forget the first time sibling murder took place at the breeding center. The mother was Saliega, known as Sali, and she was the first female ever to give birth in captivity. She was an excellent mother, and her three cubs were all doing well—until, when they were about six weeks old, a play bout between the largest cub, Brezo, and one of his sisters suddenly turned deadly serious and they began to fight fiercely. Sali seemed perplexed and tried to break it up, holding one or the other of the pair with her jaws, shaking them. But Brezo would not let go, and in the end, badly wounded himself, he killed his sister with a bite to her throat.

"We suddenly went from a happy family to an awful crisis situation with a dead cub, an injured one, and a completely stressed-out mother who would repeatedly take the third cub in her jaws and pace all over the enclosure," said Astrid.

Frantically Astrid contacted as many experts as she could. Finally she got through to Dr. Sergey Naidenko, a Russian scientist who had studied the Eurasian lynx for twenty years. And, he told her, for eighteen of those years he had recorded sibling aggression among captive lynx and he had come to think that it was normal behavior. But no one had believed him—it was always ascribed to bad management. Astrid was delighted to speak with Naidenko. "It was like finding a guru," she told me.

She asked him if he'd had success returning injured cubs to their mothers, and he said yes, 100 percent success. But, he warned, it would have to be done very carefully. At this point, Astrid had to make a tough decision: She knew Brezo needed his mother and her milk, but she also knew that the media and wildlife authorities were watching closely. What if she made a wrong decision and it led to the death of another precious lynx? She would be blamed, perhaps damaging the status of the whole breeding program. But because their goal was to return lynx to the wild, it was vital that the cubs be raised by their mothers. So she decided, with much apprehension, to take the risk.

Brezo had been away from his mother for a day and a half. First they sprinkled him with Sali's urine—she often sprayed her cubs. "We tried," said Astrid, "to cover as much as possible our human smell with Sali's own perfume." As soon as Sali saw Brezo, she began "vocalizing sounds of joy." Once he was in the enclosure, she groomed him and sprayed him and lay down so he could suckle. "Brezo was in lynx heaven," said Astrid, "and we were so happy and deeply touched by the scene that I still get chills when I recall it."

Since then, the team has broken up fights in several subsequent litters. They always occur when the cubs are about six weeks old, and for no apparent reason.

Mothers and Cubs

I was able to see firsthand how much Astrid cares for the lynx in the program. She and the head keeper, Juana Bergara, took me first to visit Esperanza, mother of the cub killed the night before. Despite this trauma—or perhaps because of it—she was clearly very, very pleased to see Astrid and Juana. Although all the lynx cubs at the center are raised with limited contact with humans, and prepared, as far as possible, for survival in the wild, Esperanza had been hand-raised and had a special relationship with people.

As we approached, wearing protective booties and rubber gloves, she gave little breathy sounds of greeting and rubbed up against the wire. She repeatedly butted the wire mesh with her head—a sign of affection, Astrid said. Clearly, she could not get enough of this attention—I had the feeling that this contact was soothing for her after the stress of the night. I heard her purring like a happy domestic cat. She had been found in 2001, Astrid told me, as an almost dead one-week-old lynx cub. She was saved by the Jerez zoo vets and hand-raised. She never saw another lynx until she was almost a year old.

To provide opportunity for the cubs to learn from their mothers, the families are kept in large outdoor enclosures, where the cubs are taught to hunt by their mothers. Rabbits, of course, are bred for this purpose. In one of the big enclosures, three cubs were playing. Their mother led them toward a handsome black rabbit, but they showed absolutely no desire to want to hurt it, nor did the rabbit show the slightest fear. It almost seemed to want to play! The keeper told me that one of the lynx refused to kill one individual rabbit that remained in the enclosure for several weeks—and thereby witnessed the quick dispatch of many others of his kind. This, of course, is the difficult part of such programs. Astrid told me that she always feels so sorry for the rabbits. It makes it worse that her son, Mario, now four years old, always wants to go and see the rabbits when he visits the facility. And he always asks to bring them home.

Astrid took me to visit two of the breeding males—they are stunningly beautiful creatures. One lay quite far away, watching us intently. The other was close to the mesh, but spat and hissed at us as we approached. He had lived in the wild until he was three years old, Astrid told me. Then he was brought to the center too badly injured to be released. As we watched him, Astrid was reminded of another injured lynx, Viciosa, who had been sent to her from Andalusia, and I remembered Miguel telling me about her when we met in Barcelona. When he had found her, by following signals from her radio collar, she'd been close to death. She had been badly injured by fighting during the breeding season, and weighed only eleven pounds instead of the average twenty-four pounds or so. Amazingly, with good care and good food, she recovered in three weeks.

When Astrid received her, she had already been saddled with the name Viciosa (which means "vicious") by Miguel's team. "But she wasn't at all vicious," Astrid told me, "she just wanted to eat and eat!" When Viciosa was released back into her territory toward the end of breeding season, she immediately coupled with a male, and nine weeks later gave birth to two cubs.

I was very impressed by Astrid's facility. There are cameras mounted to cover each outside area and others for the inside of the dens. The TV monitors are on twenty-four hours, monitored by staff or volunteers, throughout the whole year, and with particular intensity during the three months of birthing and cub rearing. All this footage is providing unique information about lynx behavior.

I was astounded by a truly unique method for collecting blood. Any attempt to anesthetize the lynx, or handle them in any way, is extremely distressing for them. A German scientist had the idea of collecting blood by means of a giant bedbug! Lynx sleep on a layer of cork at night. A small hole is cut in this, and into this space a hungry bedbug is placed. It makes a beeline—or bugline!—for the warm body and starts to suck blood. After twenty minutes (when the bug starts to digest the blood), it is removed from below the sleeping platform, and the blood removed with a syringe. The lynx sleeps on, undisturbed. And the bug can be used again! (No doubt this will cause outrage among People for the Ethical Treatment of Bedbugs!)

A Tragic Killing

Before we left, Judy and I saw the infrared footage of the night's fatal attack. It lasted eight minutes. It began when the victim, up on a ledge in the night quarters, was suddenly, for no apparent reason, attacked by her brother from the back. Then the two started fighting in earnest. The victim, from the start, went on the defensive, lying on her back and kicking with her back legs. After two minutes, the kicking stopped.

Esperanza had rushed to the scene instantly and, seizing the victim, tried to pull her away. Three times she managed to separate them, but the aggressor would not give up. Astrid had been called and was there within five minutes—but although she retrieved the cub, it was too late to save her. She had a punctured lung and several broken ribs.

After the dying cub had been removed, Esperanza behaved strangely. Every time the survivor tried to return to the den, his mother—who seemed unable to carry him in the accepted manner, by the scruff of his neck—dragged him out despite his attempts to resist. This was repeated many times. For some reason Esperanza did not want him in that den.

Later I heard from Astrid that a careful necropsy had shown that the actual lethal wounds had not been inflicted by the male sibling, as had been thought, but by the mother in her efforts to try to separate her cubs. "Esperanza," Astrid told me, "was always attentive yet rough with her cubs. Her instinct to separate these two was good, but she was captive-raised, had no lynx playmate as a cub, and thus had no chance to learn her own strength. And that," said Astrid, "was lethal."

The Future of the Lynx in the Wild

That evening Astrid, Toñe, and Javitxu drove Judy and me into the Doñana National Park lynx habitat. Of course we saw no lynx, though Javitxu told us that just the previous week he had seen a mother with three cubs playing in one of the many open clearings among the low trees.

During the drive, we discussed the many difficulties and the many problems that lie ahead—the protection of suitable habitat, for one thing. Even the national parks are not always safe. Part of Doñana National Park's buffer zone had been taken over for a golf course. Also, each year, hundreds of thousands of people make a pilgrimage to the Virgin of Rocio festival, in honor of a statuette of the Virgin Mary that once supposedly magically appeared in a tree. Unfortunately, the pilgrims pass through prime lynx habitat, right through the national park, in the middle of the breeding season. Then, too, there are more tourists coming into the area, attracted by the beautiful beaches. And as road traffic increases, so do the numbers of lynx killed on roads (at the time about 5 percent of all deaths).

Nonetheless, as we discussed over a delicious dinner in a small and friendly restaurant, there is much that is positive. For one thing, the lynx population in Doñana is now stable at about forty to fifty individuals. It is of course the number of breeding females, and the number of young born each year, that counts. During recent years, there have been ten to fifteen females.

And work has started on the construction of tunnels under the roads in the hope that the lynx will learn to use them—as animals do in other places. They are thinking about building bridges over the roads, too. Finally, and most importantly, they are working to increase the number of rabbits.

We poured the last of the Spanish red wine and raised our glasses to the restoration of the Iberian lynx and the dedicated people who are devoting their all to making a dream come true.

Postscript

Later, in the fall of 2008, I heard from Astrid that the captive breeding program was, by mid-2008, ahead of projections. There were, she said, fifty-two lynx in captivity, twenty-four of which were born in the facility. This means, said Astrid, that provided the release area is ready for them, reintroduction of captive-born lynx could take place in 2009— one year ahead of schedule. And because not one Iberian lynx had been killed in a road accident in Doñana since late 2006, it seems that the area may be suitable for reintroducing captive-born lynx.

I then heard from Miguel that the number of territorial breeding females was up to nineteen, and there were between seventeen and twenty-one new cubs alive in September 2008. While the verdict is still out as to whether or not Spain's magnificent Iberian lynx will once again have a suitable habitat that allows it to thrive in the wild—a protected area that is safe from pilgrims, golf courses, and the like—for now the news is encouraging.

John Hare, adventurer, explorer, and passionate advocate for the wild Bactrian camel, shown here with domestic Bactrian near the northern border of Tibet, surveying a sanctuary for their highly endangered wild cousins. *(Yuan Lei)*

Bactrian Camel
(Camelus bactrianus ferus)

In the Gobi Deserts of Mongolia and China, in some of the most desolate country in the world, truly wild Bactrian (two-humped) camels still live. Wild Bactrian camels were captured and domesticated about four thousand years ago. Gradually, over time, the descendants of those first domesticated herds have become genetically differentiated from their wild relatives.

Everything I know about these camels I have learned from John Hare, the man who has done more to save them than anyone else. Indeed, but for him and the Chinese and Mongolian colleagues he works with and inspires, the wild Bactrian camels would almost certainly have reached the point of no return. I first met John Hare in 1997, just before the publication of his book *The Lost Camels of Tartary*.

John was once in the British Foreign Service—one of the old brigade, tough without being burly, efficient, determined, and with a passion for adventure. Over the years, we have talked a great deal about his mission to save the Bactrian camels. When we first met, I knew no more about them than he knew about apes. I rode on a domestic Bactrian in the Kolmarden Zoo in Stockholm—just to see what it was like—and John glimpsed a few wild chimpanzees when he was serving in Nigeria. But we are both basically creatures of the wild places, and only leave them to try to save them. John has generously shared his knowledge with me, written for me something of his years with the Chinese and the Mongolians—and the wild camels.

"My desert adventures—that have, over the past twelve years, enabled me to visit the four enclaves in the Gobi in China and Mongolia where the wild Bactrian camel still survives," he wrote, "began in neither of those two countries but in Moscow. I was there, in 1992, to stage an exhibition of environmental photographs in the Polytechnic Museum. At the reception, I spotted a man in a dark suit who sported a Stalin look-alike mustache, and I asked him how he was managing to survive in lawless Moscow. For Moscow was a dangerous place at that time after both communism and law and order had collapsed. At that moment, camels and the Gobi Desert could not have been further from my mind.

"'I work for the Russian Academy of Sciences,' Professor Peter Gunin said in hesitant English. 'I lead the joint Russian/Mongolian expeditions to the Gobi Desert. That takes me away from Moscow every year and so I manage to survive.'

"'Do you ever take foreigners on your expeditions?' I asked. 'I'd give my right arm to go with you.'

"Peter Gunin stroked his bushy mustache. 'There's no market in Moscow for a foreigner's right arm,' he said with a smile. 'Even the Mafia aren't interested in them. What can you do? Are you a scientist?'

"'Unfortunately, not,' I replied, searching desperately for something relevant to say. 'I could take photos. I could come as your cameraman.'

"'My colleague, Anatoly, is coming on the next expedition as the official photographer,' Peter replied. 'Is there nothing that you can do that has a scientific background? I will have to justify your inclusion to the Academy.'

"'Do you use camels on your expeditions?' I asked. 'I've had a good deal of experience working with camels in Africa.'

"'That's it,' he cried. 'Camels! We need a camel expert. We need someone to undertake a survey of the wild Bactrian camel population in the Mongolian Gobi.'

"'I know nothing about the wild Bactrian camel,' I said. 'Nothing at all. I didn't even know there was such an animal.'

"'You will learn all about the wild Bactrian camel if you come with us,' Peter Gunin said. "He gave me a broad wink. 'Provided that you can get the foreign exchange.'

"'How much do you want?'

"'Fifteen hundred dollars, plus your air-fare.'

"'I'll try to find it,' I said without the slightest hesitation. I had no idea how I was going to get hold of it or whether I would get leave of absence from my job. I only knew that I had to go with this amiable Russian professor into the Mongolian Gobi."

As a result of that chance meeting, John has made seven expeditions into the deserts of China and Mongolia and probably knows as much as or more than anyone else about the wild camels, their habits, range, population status, and history.

Bactrian camels feed mainly on shrubs; their humps act as a rich fat store that allows them to go for long periods without food. They are also able to go for long periods without water—which is not, contrary to popular belief, stored in the humps. When water is located, they are able to drink as much as fifteen gallons at one time in order to replenish reserves they have lost. Two hundred years ago, Bactrian camels ranged across the deserts of southern Mongolia, northwestern China, and into Kazakhstan in habitats ranging from rocky mountains to plains and high sand dunes. Years of persecution have reduced the species to four small fragmented populations, three in northwest China (approximately 650) and one in Mongolia (about 450).

The Primary Enemy—Human

Their enemies are the humans who hunt them, prospect for oil in the desert sands where they struggle to survive, conduct nuclear tests in the heart of their homeland, and poison their limited grazing by using potassium cyanide in a search for gold. There could be fewer than a thousand—they are more endangered than the giant panda.

"In my quest for this timid and elusive creature," John wrote me, "I have led expeditions, four of them on domesticated Bactrian camels, into some of the most breathtakingly beautiful yet hostile country imaginable. I have traveled through forbidden areas, closed for over forty years, made the first recorded crossing of the Gashun Gobi from north to south, and been fortunate to stumble across a lost outpost of the ancient city of Lou Lan. And so, whether I was walking behind domestic camels, Bactrians or Dromedaries, or scanning the sky-line for their wild relatives, the camel has enabled me to do what I like doing best: exploring."

John has developed an immense respect for these amazing creatures, so ideally suited to their desert environment. "Recently," he told me, "I traveled with Pasha, a one-humped dromedary camel, for three and a half months across the Sahara. As I rode him day after day he became a wonderful companion. In the end he was following me about like a dog, sniffing at my trouser pocket, which held his beloved dried dates."

In 1997 John set up the Wild Camel Protection Foundation (WCPF), a registered charity in the United Kingdom, to raise funds for conservation efforts to protect the Bactrian camels in the wild. The WCPF, working with eminent Chinese scientists, persuaded the Chinese government to establish the 67,500-square-mile Arjin Shan Lop Nur Wild Camel National Nature Reserve—which is bigger than Poland and nearly the size of Texas.

It is wild and desolate desert country where few living things can survive since there is almost no water for most of the year save salty slush that bubbles up from under the ground. At one time, there was some fresh water from the spring snowmelt from the mountains. But the construction of dams and overuse of water for agriculture have more or less eliminated this, except in the mountainous area in the south of the reserve. The wild Bactrian camels have learned to survive by drinking the salty water that the domestic Bactrian would not touch—although the wild camels much prefer to drink sweet water if they can get it.

When I first met John, he was searching for funds to set up five ranger posts for this nature reserve, and I was able to persuade two generous friends, Fred Matser and Robert Schad, to donate money for three of the posts. It was not difficult—both were captivated by my description of the camels, their wild habitat, and the man who risked his life to save them. And both care passionately about conservation of the natural world.

John and I bumped into each other again in Beijing, at the China headquarters of the Jane Goodall Institute, when he was working to convene a workshop involving delegates from the governments of China and Mongolia. He needed the cooperation of both to ensure the survival of wild camels in the adjoining desert habitats of both countries. Already in 1982, the wild camels of Mongolia were protected in the Great Gobi Reserve A, and they were protected in the newly established nature reserve in China, but there was no communication between the two countries. That workshop led to a historic agreement, signed by both the Chinese and Mongolian governments, to jointly protect the wild camels on both sides of the international border. They also agreed to cooperate in a wild camel data exchange program.

However, despite these successful moves to protect the wild Bactrian camels, there is still grave concern for their future. They have been heavily hunted for their meat and hide over the centuries and are still hunted—for "sport" or because they are perceived as competing with domestic livestock for the precious water and grazing of the desert. Ironically, it was the forty-five-year stretch in which the Gashun Gobi Desert was used as a nuclear test site, when the area was strictly off limits, that provided them with their only refuge. Now, however, a gas pipeline has been built across the once forbidden desert, and it has also become infested by illegal gold miners contaminating the environment with a highly toxic potassium cyanide. Hybridization with domestic camels poses a further threat to the survival of the wild Bactrians. For all these reasons, John and the WCPF felt that it was important to start a captive wild Bactrian camel breeding program.

In 2003, the Mongolian government not only approved this idea but also generously donated a suitable area for captive breeding—Zakhyn-Us, near the Great Gobi Reserve A, where a freshwater spring provides a year-round supply. A strong fence was erected, a barn for hay storage, and three pens where captive wild camels and newborn calves can find shelter from extreme weather—important since the female gives birth during the coldest months of the year, December to April, and the Mongolian winter can be very severe with temperatures dropping to forty below Fahrenheit.

In the summer months, when the frenzy of mating has subsided and the birth season is over, the captive camels are released from the fenced area so that they can graze as a herd near their natural homeland. During this time, they are constantly supervised by a Mongolian herdsman and his family, who are employed by the WCPF to look after them. Meanwhile the grass in the penned area is given a chance to recover.

"At the end of the first three years of operation," John wrote, "seven wild Bactrian camels were born to the eleven wild females and the wild bull camel that had been caught by Mongolian herdsmen."

The last time I met John, he had some wonderful news. Recently, after a successful Edge Fellowship training course held at the Zoological Society of London, he had invited two young scientists—one a Chinese and the other a Mongolian—to spend two nights with him in the gher (a Mongolian version of a yurt) that he has built on his land in England. "There, songs were sung and the whiskey flowed, and this helped to soften prejudice and deepen friendship," he said. The two scientists are now firm friends and are in regular e-mail contact over the problems faced by the wild Bactrian camel in their respective countries. "Despite our technological wonders," said John, "it is still, and always will be, human contact that matters."

Before we parted, John gave me one of the only six winter hats that had been woven from hair shed by the Bactrian camels in the breeding program. Soon more of these hats will be available—the herdsman's wife has established a small cottage industry, selling her products through the Wild Camel Protection Foundation Web site. That soft hat is one of my treasured possessions, beside me now as I write, a symbol of hope for the future of both the people and the camels of the Chinese and Mongolian deserts.

Don Lindburg, team leader for giant pandas at the San Diego Zoo, holds Mei Sheng, the second cub to be born at his facility. Mei Sheng moved to China at age four to be part of Wolong's giant panda breeding program. (San Diego Zoo)

Giant Panda
(Ailuropoda melanoleuca)

I have never seen a panda in the wild. Few people have, even those who have spent years studying them in the field. I have seen several of those loaned out by the Chinese government to various major zoos, including the first pair sent to the Smithsonian National Zoo in Washington, DC, in 1972. More recently, I visited those in the Beijing zoo, where somewhat to my surprise the male was lounging up in the fork of a tree. Of course, I now know that they frequently climb, especially the youngsters—I just had not thought of them up among the leaves. Which is hardly surprising as most zoos have only recently begun to supply climbing opportunities for their pandas.

The home of the giant panda is in South China, in temperate mixed-broadleaf forests east of the Tibetan plateau. Although there may now be as many as sixteen hundred in the wild, their future is far from certain. One of the problems, other than habitat loss, is their diet. They are bears—yet unlike other bears, they survive only on certain kinds of bamboo. Since bamboo is not nutritious, the giant pandas have to eat very large quantities of it. So it was especially worrying when, in 1978, there was a massive die-off of bamboo in panda habitat. It was unthinkable that the giant panda, a national symbol, should become extinct. So the Chinese government sent scientists into the field to find out what was going on.

First Studies in the Wild

Professor Hu Jinchu and his colleagues erected a hut in the Wolong Natural Reserve in the Qionglai Mountains. There they were joined three years later by my old friend Dr. George Schaller, who was to collaborate with the Chinese team in a field study sponsored by the WWF. Things were difficult in China back then; after four and a half years, George felt he could contribute nothing further and left the project. Thinking back to that time, he would later write: "I was filled with creeping despair, as the panda seemed increasingly shadowed by fear of extinction."

Indeed, between 1975 and 1989 half of the habitat of the giant panda in Sichuan Province was lost due to logging and agriculture; the remaining forest was fragmented by roads and other developments. Among other things, this affects the regeneration of the bamboo, since it grows best under a forest canopy. The giant panda population became dispersed in small groups living in isolation. And that, as George wrote, is "a blueprint for extinction." The pandas were also being illegally killed by poachers.

Pan Wenshi also began working with giant pandas in the 1970s, starting his own research in the Qinling Mountains. His formal studies had been interrupted by the Cultural Revolution so that he did not start off with the academic credentials of some of the other panda researchers. Nevertheless, his project continued for thirteen years during which he and his all-Chinese team radio-collared and tracked twenty-one pandas, gaining valuable information on all aspects of their behavior.

Devra Kleiman, whose work with the golden lion tamarin is detailed in part 2, has been involved in giant panda conservation work since her first visit to China in 1978, and she got to know Pan Wenshi quite well. During one of her visits, in November 1992, Pan promised that—in honor of her fiftieth birthday—she would see her first wild panda. She set off with some of his team for a cave where a female and her cub had been denned up—but when they got there, the pandas were gone. Pan was crestfallen. Suddenly panda calls sounded across the valley, "and not only did I get to see one wild panda, but there were three—one up a tree and two on the ground," said Devra. "It was an incredibly unusual sighting, since researchers almost never saw pandas together outside the spring breeding season, especially not in November. Pan was as thrilled as I was!"

Another biologist who joined the team in the Wolong Natural Reserve in the mid-1990s was Dr. Matthew Durnin, who is now on the board of JGI-China. He told me that he only saw a giant panda in the wild once in his ten years of trudging up and down the steep densely forested slopes looking out for the telltale sign that would indicate where pandas had been—the remains of a bamboo meal and panda poop.

From time to time, students joined the team to pick up a few months of field experience. Because the research area was large, the team divided, searching in different areas and sharing information at the end of the day. "One evening," said Matt, "I returned from another panda-less day and realized, at once, that something was up—one of the students, who had only been with the project two months, not only saw a giant panda up close, but photographed it as well!" Apparently he and the Chinese researcher had come upon it when it was asleep, and on waking it had seemed groggy. They had watched for five to six minutes before it woke up fully and hurried away. The only panda Matt saw was glimpsed briefly as it moved along a distant ridge.

During his time at Wolong, Matt got to know many of the staff who were employed locally. "I learned so much from them," he told me. "They got paid very little, but they were so energetic, and seemingly enthusiastic, that you would think they chose the work—though, in fact, there were few opportunities for employment and they probably had little choice."

The caretaker Wee Pung was from a minority group, and had been working at Wolong for almost fifteen years. He seemed really proud of the reserve, and of his role as caretaker. "All that time," said Matt, "this man was keeping watch over the place, living in the woods." And even though he may have taken that work from necessity, one day Wee Pung told Matt that not once since he'd joined the project had he been able to visit his family—he just could not afford the trip. Matt assumed his family was far away on the other side of the country. In fact, he told me, "it was just a two-hour drive away." And so, of course, Matt drove him there.

Captive Breeding

The Chinese have put a lot of effort and money into captive breeding programs, but for years they had little success. Many Western scientists were invited to the Wolong breeding center to work along with the Chinese scientists for short periods—and Devra went for several months in 1982. In those days, the location was difficult to get to. They had to walk for about an hour, uphill, from the main road. And, said Devra, "They had to transport pandas by hand—two workers per panda—up the steep and slippery path, passing through two long tunnels that had been blasted out of the mountainside."

One of the problems with the captive breeding at that time, Devra told me, was a lack of understanding of panda behavior that led to inappropriate husbandry. The pandas were caged separately and had no opportunity for socializing. Even during the breeding season, males and females were seldom introduced to each other for fear of aggression. Artificial insemination was the preferred method of inducing pregnancy, and in fact there were few males capable of mating with females naturally. That was partly, Devra believes, because they had no opportunity to climb, and their legs and hindquarters were often not well devel-

Matt Durnin at work at China's Wolong Nature Reserve, checking a seven-month-old panda. *(Matt Durnin)*

oped. Sometimes the female had trouble supporting the male during copulation, and he had trouble maintaining his mounting position.

Then, in the mid- to late 1990s, the San Diego and Atlanta zoological societies, responding to requests from China, sent their scientists over to work with Chinese colleagues at Wolong. My good friend Don Lindburg, his postdoctoral student Ron Swaisgood, and Rebecca Snyder from Atlanta did a great deal of successful work there. At the same time Chinese zoos, especially the Wolong and Chengdu zoos, were also working to breed pandas.

Success

Finally, starting in 2000, births began to outnumber deaths, and from 2005 there were significant increases in the captive population. "This," said Devra, "was a direct result of a change of attitude toward managing the pandas. All the recent significant increases in the captive population numbers have come about because of better captive conditions and an increase in natural matings." Another factor was an innovative way of helping a mother panda to raise both babies when she has twins, first developed at the Chengdu Zoo Captive Breeding Center. Before this, a mother usually abandoned one of her two babies—which is not surprising, for raising two panda cubs is a lot of work. Like kittens, panda cubs cannot urinate or defecate without stimulation for several weeks—okay with one baby, extremely difficult with two. Now, however, a human caretaker gives the mother a helping hand: The twins are rotated, and while the mother cares for one the human surrogate takes over the other. As a result of all this, in 2008 there was a 95 percent survival rate in infants born at Wolong, compared with 50 percent twenty years before.

First Months of a Giant Panda Cub

Recently I had dinner with my old friend Harry Schwammer, director of Zoo Vienna, which is also involved in the giant panda captive breeding program. He told me that they recently experienced their first panda birth. Head keeper Eveline Dungl told me how the mother, Yang Yang, had built a branch-lined den in her outside enclosure, but subsequently moved inside to the specially prepared nesting box. Two mornings later, Eveline heard squeaks "that definitely did not come from Yang Yang."

Yang Yang was an excellent mother, and not until the baby Fu Long was two and a half months old did she leave the den for a few hours at a time to feed outside. "Now, at the age of almost one year," Eveline wrote to me, "Fu Long is already quite self-confident in exploring his surroundings. Even though he still mainly drinks breast milk, he is very interested in bamboo. And he also likes to try leaves or branches of other plants. There is no tree in the enclosure that he has not already climbed, no platform he hasn't napped on."

Harry Schwammer and his staff are engaged in discussions with Chinese scientists about the program to reintroduce giant pandas to the wild. Harry and others believe that it will be important to rear cubs with minimum contact with their human caretakers. But as we shall see, there are many other challenges.

Problems with Reintroduction to the Wild

The idea of reintroduction to the wild in China was vetoed in 1991, and again in 1997 and 2000, on the grounds that there was insufficient knowledge, especially with regard to the status of wild pandas and their habitat. It was also felt that there were insufficient funds for such a long-term project. Finally, none of the current regimes of captive breeding could provide suitable candidates. However, in 2006 Xiang Xiang, a young male born at the Wolong breeding center, was released into the Wolong Natural Reserve. In the documentary film I saw, he appeared to be doing all right. His keeper showed him how to choose good bamboo, and readings from his radio collar showed that he sometimes made journeys of more than five miles—after which he always returned to the release site. However, this seemingly good start ended in tragedy when he was apparently attacked and wounded by the original panda residents. And although he recovered from those injuries, he was attacked again and died of his wounds.

Tourism and Awareness

Today many Chinese schools teach their pupils about giant panda behavior and conservation, especially in Chengdu in Sichuan Province, where local pride in the panda is strong. And indeed, the giant panda has put Chengdu on the tourist map. It's the gateway city for visiting the Wolong Giant Panda Reserve Centre, which gives talks and shows films to visitors, and allows them to play with small panda cubs. What a shock for the group of American tourists that were enjoying this experience when the terrible 2008 earthquake devastated the mountains of Sichuan Province. A *New York Times* article found the group full of praise for the "kindness and heroism" of the panda keepers, who helped them reach the road. "Those keepers were risking their lives," said one visitor. "There was nothing safe about any of it." And once all the visitors were secure, the keepers hurried back and rescued all thirteen baby pandas, carrying them tucked under their arms as they negotiated the dangerous rock-strewn route. During the quake, most of the enclosures were destroyed; one panda was killed, two were injured, and six escaped (four of which were later captured).

Of course there was immediate concern and anguish for the thousands of people who were affected, particularly the children who perished in the cheaply constructed schools. (All ten schools with JGI Roots & Shoots groups were affected. Most of the teachers and students lost their homes, and many lost family members. Their school buildings are mostly either collapsed or unusable. One young boy was killed.)

There was also national and international concern for the wild pandas, most of whom are living in the forty-four nature reserves in the Sichuan mountains. Dr. Lu Zhi, a leading panda expert and China country director for Conservation International, said researchers were trying to find out how the wild pandas had been affected, even as they were helping with the human tragedy.

"The Panda's Day Is Now"

During the 1990s, there was a change in China's conservation policy when, as a result of massive flooding in the Yangtze River Basin, the government imposed a ban on commercial logging and launched a vast reforestation effort on the steep hillsides—where clear-cutting had removed the cover necessary for protection of the watersheds. Luckily for the giant panda, much of that area fell within its range. For the Chinese, the giant panda is a national treasure, and suddenly it seemed possible to set aside new reserves for them. Most recently, in 2006, the government expressed even stronger support for the protection of the panda's habitat when the provincial governments of Sichuan and Gansu agreed to expand and connect scattered nature reserves in the Minshan mountain range, home to about half of the approximately 1,590 wild giant pandas believed to live there.

Over the years, conferences to discuss panda conservation have been held in Berlin (1984), Tokyo (1986), Hangzhou, China (1988), and Washington, DC (1991). In 2000, the San Diego Zoological Society brought together scientists from China, Europe, and North America to discuss current understanding of the giant panda. Known as Panda 2000, this conference created new collaborations and new friendships and provided a great deal of new information, which is presented in a major volume, *Giant Panda: Biology and Conservation*. In his foreword, Don Lindburg wrote: "Perhaps the clearest consensus drawn from this event was that *the panda's day is now*."

And George Schaller, so pessimistic when he left China in the 1980s, wrote in his introduction to the book: "The prospects for saving the giant panda are today unequaled."

THE BIRTH OF A PANDA: A SYMBOL OF NEW COOPERATION

A few months ago, I met up with my friend Donald Lindburg in California to discuss his years of involvement with the panda breeding programs in Wolong and San Diego. He told me of the birth he had witnessed, and I asked him to send me an account. It was in San Diego in 1999, the first birth since those at the National Zoological Park in the late 1980s.

Bai Yun's pregnancy had been going well. "A veterinarian had recently confirmed the presence of a fetus in Bai Yun's womb via ultrasound," wrote Don, "and her hormone profile led to predictions of a birth within days. Now, the twenty-four-hour watch has begun, and as a video monitor portrays mother in her birth den, the staff waits in hushed silence. There was plenty of evidence to indicate that at this crucial moment in time, something could go wrong, very wrong, giving [rise] to mixed emotions of hope and anxiety.

"Early in the day, the first signs of labor were evident. As the pace of Bai Yun's straining increased, suddenly there was a scratchy-sounding wail, a sound never before heard by the eager watchers. Immediately, two staff members who were on rotation from the Wolong Centre in China—and who had witnessed previous births—gave the thumbs-up sign.

"All eyes were glued to the video screen when first-time mother Bai Yun bent over and retrieved her seconds-old cub from the floor of the den. She placed it on her expansive ventrum and began to lick it vigorously. Soon, there was a new kind of sound from the cub—a sound we would later call a contentment vocalization—as it dozed off for its first postpartum nap.

"The air was electric with excitement. Everyone in the room wanted to shout and clap, but showed great restraint out of fear that the mother in the nearby den might be disturbed. In the days immediately following, *Good Morning America* and the *Today* show, as well as local media, would check in for the latest word on this rare event. By diplomatic message, sent secretly from China to its Consulate in Los Angeles, at one hundred days of age this new baby would be named Hua Mei, meaning 'China–USA.'

"The symbolism was clear. A single birth will not save the species—but now a new direction in its conservation had been noted."

The pygmy hog population was once down to only a few survivors dwelling in Manas National Park in India. Many dedicated individuals helped to restore this unique and highly intelligent animal through captive breeding. *(Goutam Narayan)*

Pygmy Hog
(Porcula salvania)

I've always loved the pig family. The first animal I "habituated" was a saddleback named (by me) Grunter. He was in a field with about ten others. I took him my apple core after lunch every day during a summer holiday and eventually he let me scratch his back. What a triumph!

One of my treasured memories from my years in Gombe was the time when a sounder of bushpigs came upon me as I sat very still in the forest. They could not make me out, stared and sniffed the air, came closer—until I was surrounded. One gave a snorting alarm call and they ran off a few yards, but returned to stare in silence. Finally they moved on, rustling through the leaves eating fallen mbula fruits. I've spent time, also, watching another member of the porcine family: warthogs on the Serengeti plains, grazing on their bent knees, running with tails straight up, competing with each other for the best dens in which to sleep for the night. And I've glimpsed wild boar while driving at night in Germany, Hungary, and the Czech Republic.

When I first saw pygmy hogs—a pair in Zoo Zürich—I could scarcely believe my eyes. A pig, measuring at the most one foot in height, and weighing a maximum of twenty pounds! I was sure I was looking at two juveniles—yet they were perfect little adults, dark brown with coarse hair, short stubby legs, and a minute tail. There was a slight crest on the forehead and nape of neck, and a tapering snout. I could just see the canines peeking from the mouth of the male.

The man who first described these diminutive beings in 1847, B. H. (Brian Houghton) Hodgson, must have been very amazed. He reckoned they were a different species, and although other scientists later declared the pygmy hog to be a relative of the wild boar, Hodgson was eventually proved right. Recent genetic investigations indicate that pygmy hogs belong to a unique genus, with no close relatives.

They live in tall dense grassland where they eat an omnivorous diet of roots, tubers, various invertebrates, eggs, and so on, feeding during the day unless it is very hot. They make quite elaborate nests, often digging a trough with snout and hooves, piling up soil around the edges, lining it with grass they bend down on each side, and bringing more in their mouths to make a roof. A couple of females and their young may share one nest, while the adult males, who are usually solitary, make their own. Their main predators are the python and dhole (also known as the Asiatic wild dog). And, of course, humans. (For those who like trivia, let me reveal that the pygmy hog is sole host to the pygmy hog sucking louse *[Haematopinus oliveri]*, a louse that is classified as critically endangered and named after William Oliver, the chairperson of the IUCN Pigs, Peccaries and Hippos Specialist Group.)

I had no idea, when I saw that pair in Zürich, that pygmy hogs were so endangered. At one time they ranged from Bhutan to Northern India and Nepal. But their numbers have been declining in the wild throughout the past century due to a combination of factors: expansion of human population in the Brahmaputra flood plain region, overgrazing, commercial forestry and flood control programs, taking of grass for thatching, and, especially, burning. As a result, by the late 1950s it was believed that pygmy hogs had become extinct, and they were so listed in 1961.

Ten years later J. Tessier-Yandell, a tea planter from Assam, visited Gerald Durrell at his zoo in Jersey, England, and asked if there was any special animal in Assam that he was interested in. Laughing, Durrell said, "Yes, get me a pygmy hog." And he did! He found four that were being sold in a tea garden market! They had been captured hiding in a plantation when a small forest patch nearby was burned. It was hoped they would breed, but there were no professionals there to advise and nothing came of it, although several more wild hogs were acquired. Clearly the pygmy hog was not extinct and Durrell, delighted, made plans for a captive breeding program and acquired funding for field research.

William Oliver, at that time scientific officer for the Gerald Durrell Jersey Zoo, organized extensive field surveys in the mid-1970s, and concluded that the only remaining small groups of the pygmy hog were in Assam, in the plains to the south of the Himalayas. There were no more than a thousand individuals, and habitat destruction was continuing.

It was in 1977 that the two pygmy hogs that I met were sent to the zoo in Zürich. At first all went well: The sow farrowed and delivered healthy piglets. But then she died in an "accident." The piglets remained healthy, but the only female among them was, unfortunately, left with her father and brothers. She was only one year old when she became pregnant (far too young) and she died in childbirth. That hope for captive breeding was thus ended. The only other pygmy hogs sent to Europe had gone to London Zoo in 1898 where both members of the pair had died without raising young.

In 1996, with a grant from the EU, the Durrell Wildlife Conservation Trust (then the Jersey Wildlife Preservation Trust) got permission to start a captive breeding program in Guwahati (the capital of Assam), and six pygmy hogs were captured from the last surviving population of the species in Manas National Park.

Early in 2008, on the advice of Gerald Durrell's wife, Lee, I called Goutam Narayan, who heads up the program. The voice that traveled to me from India was warm, and he was generous with his time. He explained that, with help from Parag Deka, the excellent veterinarian on staff who has been with them from the start, the breeding program was going well. "We followed established breeding guidelines—and common sense," he said. Usually four or five young are born once a year. They weigh barely five or six ounces at birth, grayish pink at first, then develop faint yellow stripes by the second week. They live up to eight years in the wild but can reach ten years in captivity.

I asked Goutam if he could share any stories from his long years with the project. He told me about a local forest guard in Manas who had rescued a young hog that he had found, half frozen and almost dead, floating down a river on a cold day in October 2002. Veterinarian Parag Deka rushed to Manas and tried his best to revive the hoglet. As its condition deteriorated, it was brought to the breeding center in Guwahati where, against all odds, the little male pig miraculously recovered. He has proved a valuable addition to the breeding program, bringing new genes from the wild, and he has sired several litters during the last six years.

"From the six original individuals," Goutam told me, "we now have about eighty individuals, divided between two centers." He said that the hogs were ready for release into the wild, "but the problem is the continuing exploitation of the environment." I could hear the frustration in his voice. The pygmy hog, he explained, is "a good indicator species"—very sensitive to disturbances in composition of the herbs and other plants in the grass. And then he went on to emphasize that "they *must* have grass for their nests." They hide in their nests and get protection from the heat and cold. "They must have grass all the year round," he reported, "all of them."

Meanwhile the Durrell Wildlife Conservation Trust, alongside the Pygmy Hog Conservation Program, had been working under the guidance of William Oliver and in partnership with the Assam Forest Department, to draw up plans for the long-term management of the species, and to find a suitable site for release. And in the spring of 2008, just four months after I spoke with Goutam, three groups of pygmy hogs, sixteen individuals in all (seven males and nine females), were taken to a facility near Nameri National Park with the goal of creating a second population of the species in the wild. There, with minimal human contact, they lived for five months in pre-release enclosures designed to replicate natural grassland habitat, getting ready for life in the wild.

At last the day came when they were moved to their final destination, the Sonai Rupai Wildlife Sanctuary, 110 miles northeast of Guwahati. After two weeks in prepared enclosures the doors were opened and they were free to leave. Their movements have been followed by direct observation at bait stations and examination of droppings and nests. Goutam told me in a recent e-mail that most of them are doing well and one of the females has even farrowed in the wild.

A major education outreach program has been initiated in the villages in the area, as it is certain that, without the cooperation of the local people, these little pigs will have no chance of surviving in the wild. At the time of writing, two other potential release sites have been found in Assam, in the Nameri and Orang National Parks. On one of my trips to India I am determined to accept Goutam's invitation to go and meet these enchanting little pygmy hogs and the dedicated people who are working so hard to save them.

Northern Bald Ibis or Waldrapp
(Geronticus eremita)

In February 2008, I met Rubio, one of thirty-two northern bald ibis or "waldrapp" that live at the Konrad Lorenz Institute in Grunau, Austria. These birds are about twenty-eight inches in length with the long curved bill that characterizes all ibis. They have a distinctive fringe of plumage around the nape of their necks, but their heads are bare with no facial or crown feathers except during the juvenile stage. I had hoped to sit on the grass while they flew freely around us, as they normally do, but unfortunately all were temporarily confined because the rate of predation had been unusually high.

I went into the huge flight aviary with one of the keepers and Dr. Fritz Johannes, who is in charge of the project. Seen close up, they were beautiful, for we were lucky with the weather: The cold winter sun brought out the glorious iridescent sheen on their almost black plumage, and shone on their long pink bills and pink legs. The juveniles, whose feathers are bronze, had not yet lost their feathered caps.

At first the birds preferred to take mealworms from their keeper and from Fritz, but then Rubio decided I was okay, too, and transferred from Fritz's shoulder to mine. Having consumed an inordinate number of mealworms, he began the serious business of grooming me. What really amazed me was how warm his beak felt, and how delicately and gently he used it as he preened my hair. He also made attempts to probe into my ears and nostrils—I must admit I was not too thrilled about that!

Eventually he was persuaded to return to his keeper—but not before he marked me with white liquid down the back of my jacket. This, of course, is a sign of good luck, so I tried to feel grateful!

I had the huge good fortune of visiting Rubio, one of the hand-reared northern bald ibis in Austria, who are being taught to migrate south for the winter by following ultralights. *(Markus Unsöld)*

I was there, with the team from JGI-Austria, to learn about the attempt to teach the waldrapp to migrate from Austria to the south of Italy. In the flight aviary next to Rubio's were the birds that would take part in the spring migration over the Alps.

Extinction in Europe

This ibis once ranged in arid mountainous regions from Southern Europe to northwestern Africa and the Middle East. Today, however, it is an extremely rare species, extinct throughout nearly all of its range as a result of pesticide use, habitat loss, and hunting for its tasty flesh. The last waldrapp disappeared from Europe in the seventeenth century. In the 1980s, after all the individuals from the last remaining wild colony in Turkey had been captured for captive breeding, it was thought that the species was extinct in the Middle East.

Between 1950 and the end of the 1980s, the last *migratory* colonies in the Moroccan mountains vanished. Fortunately, however, birds from that colony had been captured during the 1960s for exhibition in European zoos, and they became the founder individuals for an international zoo breeding program. I saw descendants of those original captives in Innsbruck, where they have been bred for forty years.

By 2000, it was believed that only one colony of about eighty-five breeding pairs of (*nonmigratory*) bald ibis remained in the wild, in the Souss Massa National Park in Morocco. But then, to ornithologists' surprise and delight, a tiny group was located in the Syrian desert. There were only seven birds, but there were three nests, and they were raising young—seven fledged in 2003.

A Human-Led Migration

The (usually) free-flying breeding colony that I visited in Austria was established in 1997. Waldrapp can survive well in the Austrian Alps during the summer, feeding on insects and other invertebrates, but they cannot endure the winter months in the wild. To create a self-sustaining population, then, it would be necessary that they learn to migrate— as in the past—to warmer climes. And so a feasibility study (based on the pioneering work with Canada geese and whooping cranes described in the last section) was planned to find out whether the bald ibis could also learn to follow ultralight planes—or trikes, as they're called—on a migration route over the Alps to Tuscany in Italy.

Unlike the whooping cranes—which, as we have seen, are raised by caretakers wearing strange white gowns to prevent them from imprinting on humans, these ibis are hand-reared and bonded closely with their caretakers. They are exposed to the sound of the trikes, and the foster parent—Fritz's wife, Angelika—wears the helmet she will don when flying the plane.

During training, the birds initially flew too far away from the trike, despite Angelika's constant calling. But their performance gradually improved, and the first successful migration started out on August 17, 2004, with nine waldrapp following two trikes. Just over two months later, on September 22, the trikes arrived, along with seven of the waldrapp, at the chosen wintering ground, Laguna di Orbetello, a WWF nature reserve in southern Tuscany. (The other two waldrapp failed to make the journey on their own and were brought along in boxes.)

The following year, using a different trike (with old-fashioned wings and more powerful engine), the same route was followed and, with fewer stopovers, took only twenty-two days, from August 18 to September 8. Because this trike could fly at a lower speed, the birds were able to follow more closely so that the whole operation went more smoothly.

During the winter of 2004–2005, following their arrival in Tuscany, the young birds stayed close to the night roost, seldom venturing much more than half a mile. However, when summer came they began to go on longer flights—up to twelve miles—before returning. And some were seen along the migration route, heading for Austria. After some weeks they returned to Tuscany, but it seemed that the instinct to migrate was still present, and Johannes, Angelika, and the rest of the team were much encouraged.

In spring 2006, all the birds who had followed trikes from Austria to Tuscany in 2004 went on long flights while those from a second successful migration in 2005 remained in the wintering area. Thus it seems that as they get older, they are more likely to leave for the breeding grounds in Austria at the time of the spring migration, and that this is genetically programmed.

That 2006 spring was an exciting time for Fritz, Angelika, and the rest of the team. They received a number of reports of sightings, mainly from bird-watchers and hunters, of individual waldrapp that had gone on these long flights—some as far as three hundred miles. Most of them had retraced the route they had been shown by humans. A few were way off course. In some cases, this may have been because, during their human-led journey, they had been carried part of the way in their boxes (the few that were not following the plane and had to be collected); thus their "memory" of the journey was incomplete.

Finally, in spring 2007—success! Four of the waldrapp who had been led south from Grunau in 2004 had become sexually mature, and to everyone's delight they flew to Austria. These were the female Aurelia and the males Speedy, Bobby, and Medea. They all returned to Grunau safely—"the first complete migration circle of the birds independent of humans," Fritz told me proudly. Places chosen for stopovers were not necessarily the same as those where they'd stopped during the human-led migration, but seemed determined by the type of habitat. Once back, Aurelia bonded with Speedy: They bred and raised three offspring.

The 2007 autumn migration to the wintering grounds in Tuscany started with some confusion, as the seventeen migratory birds got mixed up with the almost forty free-flying birds at Austria's Konrad Lorenz Institute. There they lost their motivation to migrate, preferring to stay with the others instead of heading south. It was finally decided to catch the confused birds and release them about thirty-five miles southward. One adult and one of Aurelia's juveniles evaded capture and stayed in Grunau, but four adults, including Aurelia and Speedy with their remaining two offspring, headed south as hoped.

Some of the birds were fitted with a GPS data logger. This stores the position of the bird every five minutes and can be downloaded, once the bird is in range, so that researchers can reconstruct the flight path in detail. The data showed that they had exactly followed the route along which they had been led in 2004. On September 15, Medea, Bobby, and Aurelia with her offspring—but not Speedy—were seen in Osoppo, northern Italy. Five days later—one day after the parallel human-led migration ended up at Laguna di Orbetello—Aurelia (without her offspring) and Medea also arrived in Tuscany. Bobby arrived two weeks later, but the two juveniles have not been seen since.

And what of Speedy? His story is fascinating. Even during the first migration, he flew separately from the others. In spring 2007, he started alone, flying to northern Italy, then on to Slovenia and from there to Austria. Not stopping, he continued on to Styria, near Leoben, then farther northeast until he was close to Vienna. There he turned back to Styria, where—miraculously—he met up with Aurelia and Medea. He and Aurelia then flew together to Grunau.

Then in the autumn, when the group set off to fly back to Tuscany, Speedy once again separated from the group. This time he had been

Flight formation of the ibis in northern Italy, as seen from an ultralight. *(Markus Unsöld)*

All seventeen waldrapp at the start of the 2007 autumn migration. *(Markus Unsöld)*

selected to carry a satellite transmitter instead of GPS. This technology only stores some positions every third day, but the advantage is that the researchers get these positions in real time.

Unfortunately, the device did not work—only transmitting one position on September 18. But this was a very interesting data point, because it was exactly on the flight route—which had been reconstructed from Speedy's spring GPS data—that he had followed in the spring. In other words, he was retracing his own unique flight path back to Tuscany. "We got no further satellite positions and also no sight report," said Johannes. Speedy seems to have disappeared. "Nevertheless," he told me, "the migration of these adult birds was a great success for our project. Aurelia, Medea, and Bobby are the first free-living, independent, migratory northern bald ibis in Europe after about four hundred years! That's a great motivation for us."

One Step Closer to Success

As I sit writing this chapter in Bournemouth, in August 2008, I receive an e-mail from Johannes in Slovenia. He tells me they are trying a new route, as a result of the problems of the previous year. Now they are leading the young ibis *around* the Alps instead of crossing them—and "it is fantastic till now," he writes. The birds have performed well, flying more than sixty miles per day, much farther than in previous years.

And there is also news of the six older sexually mature birds that had learned to follow the trikes. In April, they migrated northward from Italy to Austria. As during the previous year, they ended up in Styria, about fifty miles from their breeding place. All six were then taken to a small village in northern Italy close to the original migration route, where a suitable aviary had been prepared. One pair has bred and successfully raised two birds. The aviary was opened in July. So far the birds have remained close by, but Johannes expects that all eight "will start the migration within the next ten days." If the group reaches Tuscany—"and there is a good and realistic chance of that," says Johannes—"then we could definitely show that human-led migration is a suitable methodological tool for the introduction of independent, migratory groups of northern bald ibis." That will be a great success for the team.

Johannes ends by telling me of the plan to start a new project in 2009 in the Moroccan Atlas region, one of the most important breeding areas for northern bald ibis until the 1980s. The first step will be to explore the food availability in a region in the northern Atlas with some hand-raised ibis.

I think of him, there with his wife and the rest of the team, getting ready for the next flight on the way to Tuscany. And if I close my eyes, I can imagine myself back in the aviary in Austria, sitting with Johannes and Rubio. There I fell in love with these endearing birds, so totally different from whooping cranes. I can almost feel the gentle touch of Rubio's warm pink beak as he groomed me. When it had been time to leave, I had given him the last of the mealworms and, reluctantly, left the colony—to continue with my own, never-ending migration around the planet.

Rod Sayler and Lisa Shipley are working tirelessly to ensure the pygmy rabbit's survival. Shown here at the endangered species breeding facility at Washington State University, Pullman. *(Shelly Hanks)*

Columbia Basin Pygmy Rabbit
(Brachylagus idahoensis)

In 2007, my tour took me to Washington State University (WSU) in Pullman for a lecture. It was there that I heard about the Columbia Basin pygmy rabbit, and the efforts being made to save it from extinction. Once you have seen one, you fall in love—a perfect little rabbit, the smallest in North America. An adult fit easily onto the palm of my hand. Childhood images of Peter Rabbit and his siblings, Flopsy, Mopsy, and Cottontail, thronged my mind. I was hooked!

The Columbia Basin population has been isolated from other pygmy rabbits for thousands of years, and is genetically differentiated from those found in Idaho, Oregon, Montana, Nevada, and California. They are specialist feeders, able to live on sagebrush in arid western US rangelands. They need tall, dense sagebrush plants for protection as well as food, and soils that are deep enough for the construction of a burrow system. They are one of only two North American rabbits that actually dig their own burrows.

Starting in the early 1990s, numbers of pygmy rabbits in Washington State declined following loss of habitat and fragmentation of the remaining sagebrush ecosystems as ever more land was taken over by farms, ranches, and urban development. In 1999, the Washington Department of Fish and Wildlife asked Dr. Rod Sayler and his colleague, Dr. Lisa Shipley, if they would help them conduct studies on the declining populations. At the time, Rod and Lisa were working on assessing the influence of cattle grazing in sagebrush habitats known to be important to pygmy rabbits. These studies had barely started when it was discovered that the largest remaining pygmy rabbit population had just suffered a major crash—possibly due to disease. Probably fewer than thirty individuals remained. USFWS gave these rabbits a temporary emergency endangered listing in 2001 with a final ruling to cement the listing in March 2003. At this time, it was decided to start a captive breeding program with the goal of subsequently releasing them back into the wild.

Sixteen rabbits were captured and sent to three facilities for captive breeding. If any were left in the wild, they soon vanished. Oregon Zoo had already started breeding the non-endangered Idaho pygmy rabbits in order to experiment on the best procedures before trying with the precious remnants of the Columbia Basin population. Rod and Lisa, heading the captive breeding program at Washington State University, found that it was necessary to house the rabbits alone, except for mating, because of high levels of aggression. Much was learned from observing the rabbits at night through remote cameras and infrared lights.

It soon became apparent that, unlike Idaho rabbits, the Washington individuals had much lower reproductive success—fewer kits per female, lower kit growth rates, and some bone deformities. And all three sites struggled with disease and parasites. Eventually it was concluded that this was partly caused by inbreeding depression resulting from reduced genetic diversity in the small captive population. Every time a genetically important rabbit died, it meant that more diversity was lost and the chances for long-term viability of the tiny remaining population were reduced. Eventually in 2003, the USFWS Recovery Team regretfully came to the conclusion that the only way to improve the reproductive fitness and thus save the last Columbia Basin pygmy rabbits was to allow some of them to mate with Idaho rabbits. This, as had been hoped, considerably boosted the breeding success and the health of the hybrid offspring.

Eventually, after six years, it seemed realistic to make plans to reintroduce some of the Washington rabbits into the wild, and once again Idaho rabbits paved the way. Forty-two captive-bred Idahos, equipped with radio collars, were released into the wild in Idaho. They did well, and following the release at least two surviving females gave birth.

The Story of Grasshopper

My visit to WSU happened to be just before the first twenty captive-bred Columbia Basin pygmy rabbits were due to be released in eastern

Washington, a hundred miles from the university, on March 13, 2007. Each was fitted with a little radio collar so that its movements could be monitored. Everyone was excited and hopeful, but everyone knew there was no guarantee of success. I met Len Zeoli, a mature PhD student, who would be studying the rabbits' adaptation to the wild. And I met Grasshopper, one of the male rabbits due to be released. What an utterly adorable little rabbit he was—I was saddened that he would have to carry a radio collar. Tiny though it was, he was tiny, too.

Of course, I was eager to hear how the release went. The report came back from Len that things had gone well, and the rabbits had been "very rabbit-like." But there were unexpected problems—almost half the rabbits dispersed from the release area, traveling off presumably in search of new homes or mates. That did not happen in the test reintroduction in Idaho. In addition, losses to predators (coyotes, raptors) were high.

I asked specifically about Grasshopper. I was told that he, together with his brother Ant, were among the eight males who moved beyond the range of the telemetry equipment, which covers three-quarters of a mile. Eventually they were both located—just a few hundred yards from the field station where Len was staying. They had somehow made it across three and a half miles of inhospitable and sometimes rocky terrain. Knowing that they would not be able to last very long, Len captured both Grasshopper and Ant, and they were returned to captivity.

"Throughout the whole reintroduction program, everyone was pretty discouraged," Rod told me, "but then something amazing happened that restored a little bit of hope." One day when Len was keeping watch, a pygmy rabbit kit suddenly popped out of one of the artificial burrows they had installed. It sat there looking at him and he was able to get close-up photos. "We saw the kit periodically throughout the remainder of the summer," said Len, "and it became famous in a photo widely published in a news release."

The photo proved that captive-bred pygmy rabbits would breed in the wild in their first breeding season—if they could escape predators long enough and readapt to the arid sagebrush habitat. "By the end of summer," Len said, "the remaining two released rabbits were taken by predators, and we terminated the field study for 2007. Everyone had hoped for greater success, but at least they have learned a lot that will help them plan better in the future."

Rod and Len, I hear, have completed population modeling studies and concluded that the captive breeding population needs to be at least doubled so that more can be released to the wild. Since the first litters of the year usually die, perhaps because of cold wet soil, research associate Becky Elias is setting up breeding pens in a greenhouse. They are building much larger, more natural pens so that the rabbits will be better adapted to a natural environment. There are plans to release the next batch of young rabbits into a temporary enclosure at the release site to protect them from predators while they adjust to living in the wild. Sadly, I heard, both Grasshopper and Ant died before they could be released again, with a better chance of making it—but a new batch of young kits has been produced in captivity for future reintroductions.

Rod Sayler summed it all up: "We're definitely not over the hump in terms of restoring this endangered species back on the landscape—there are big challenges ahead for this little rabbit. But we still have hope! We learned a lot from last year's release, and we're not giving up."

My best wishes are with you all, the humans involved and all of the enchanting little rabbits.

Attwater's Prairie Chicken
(*Tympanuchus cupido attwateri*)

The Attwater's prairie chicken, like the less rare greater prairie chicken, is a lekking species. That is, the males gather together on a carefully selected patch of short grass, or bare ground. On either side of their necks are bright orange air sacs that, when inflated, enable the males to utter booming sounds as they challenge one another. Females, attracted by the sound, gather at the lek to choose a mate.

The prairie chickens are grouse, ground-nesting birds about seventeen to eighteen inches long and weighing about one and a half to two pounds. The Attwater's plumage is striped with narrow vertical bars of dark brown and buff white, and the male has elongated feathers (called pinnae) on his head that stand up like little ears. They are smaller and lack the feathering extending to the feet that characterizes the greater prairie chicken. The Attwater's is also a bit darker in color—tawnier on the top, with a pronounced chestnut-toned neck.

I have never seen an Attwater's prairie chicken, let alone seen a mating display. But I have watched greater prairie chickens during the mating season in the Sand Hills of Nebraska. Tom Mangelsen and I arrived before first light, hoping we would see the spectacular (but also comic) display of the male.

The first prairie chickens appeared when it was not light enough to make out their colors, but soon the rising sun lit the brown-barred body feathers, the black short-tail feathers, and the brilliant orange-red of the air sac and eye combs. We had the most amazing show as more and more cocks gathered on the lek. The individual closest to us seemed to be the dominant one. Every so often, he started his display—booming,

A male Attwater's prairie chicken challenging other males during the breeding season. *(Grady Allen)*

lowered half-stretched wings, raised tail, and inflated air sacs. This was accompanied by very rapid stamping of the feet. Once in a while, one cock would start running toward another with fast little steps, head lowered, and wings stretched out. When he got close, he stopped and the two stared at each other before leaping up and down and hitting out at each other with their feet. After they had repeated this challenge several times, one would run off, presumably defeated.

Eventually a hen appeared—which caused an intensification of the displays and skirmishes. The small female seemed totally indifferent to all this activity as she moved about in the lek. (We were told that this was not the peak of the breeding season—otherwise more hens would have arrived and things would have heated up.) The show lasted about two hours, and then the birds wandered off into the vegetation. What an enchanting morning. I decided that God must have created the prairie chicken so He could have a good laugh anytime He wanted during the three-month season of the lek! It is said that some of the dances of the North American Plains Indians, particularly the Lakota, are based on this display—I would certainly love to see one!

At one time the Attwater's prairie chicken was found throughout some six million acres of the tallgrass prairie ecosystem, from the Gulf Coast of Texas north to Louisiana and inland for about seventy-five miles. The windswept prairies were rich in biodiversity then, with many varieties of grasses. But in a sequence of events we are all too familiar with, more and more of this pristine land was taken over by human development and farming, and bush invaded the grassland when fires were suppressed. Year by year, the prairie chickens vanished: By 1919 they had gone from Louisiana, and by 1937 fewer than nine thousand remained in Texas. In 1967, Attwater's prairie chicken was listed as endangered, and six years later the Endangered Species Act of 1973 gave added protection.

Today less than 1 percent of the original prairie once occupied by Attwater's prairie chickens remains, much of it so fragmented that remnant pockets are too small to sustain viable breeding populations. Fortunately a refuge was established in the mid-1960s when WWF bought an area of about thirty-five hundred acres. In 1972, it was transferred to the USFWS, and today the Attwater's Prairie Chicken National Wildlife Refuge, sixty miles west of Houston, is more than three times its original size and comprises one of the largest remnants of coastal prairie habitat in southeast Texas. The only groups of Attwater's prairie chickens in the wild today, other than those in the refuge, live on a tiny piece of land near Texas City.

The recovery plan for these birds calls for the establishment of three geographically separate viable populations—a total of about five thousand individual birds. To reach this goal, the USFWS first developed an active public outreach and education program to garner support for the birds; second, it is continuing active research; and third, it's cooperating with government agencies and private landowners to manage prairie chicken habitat. A captive breeding program with the goal of reintroducing the prairie chickens into the wild was started in the early 1990s.

The first chicks were hatched at Fossil Rim Wildlife Center, Texas, in 1992; other organizations, such as Texas A&M University and several zoos, are taking part. Once captive-raised chicks become capable of independent survival, they are sent to a planned release site where their health is checked and they are fitted with radio transmitters. For two weeks, they are cared for in acclimation pens; then they are released into their natural environment. It seems that they are genetically programmed to adapt almost immediately to life in the tallgrass prairie. In other words, once free, they behave to the manor born.

Locals Offering Safe Harbor

In 2007, a new safe harbor agreement between the Coastal Prairie Coalition of the Grazing Lands Conservation Initiative and the USFWS was finalized to help private landowners to be part of the conservation effort to restore and maintain coastal prairie habitat. In August, thirty captive-bred juveniles from various facilities were released onto private ranch-land in Goliad County, Texas, a stretch of prairie that has been kept intact by the same family since the mid-1800s. It was a milestone event, the first-ever release onto private land, and other chicks will be released there throughout 2008 and 2009. It is hoped that many more landowners will participate. Other captive-bred birds have been released on Texas Nature Conservancy property near Texas City and the Attwater Prairie Chicken National Wildlife Refuge near Eagle Lake, Texas, where Terry Rossignol is refuge manager.

Throughout 2007, staff at the refuge worked hard to increase numbers of chicks hatched in the wild. Out of a total of eighteen nests (two of which were destroyed, but remade), twelve were successful, and seventy-seven chicks made it to two weeks of age. During these first weeks, they are very vulnerable—to predation, flooding, and starvation. It is, therefore, desperately necessary to keep as many as possible alive during this time. It was decided to ask for help from volunteers. Forty-three individuals stepped forward—Fish and Wildlife employees from across the region, a school group, a master naturalist group, and various others. Their job was to assist in collecting insects for the chicks and their mothers.

Each volunteer, armed with a large canvas net and some plastic bags, was sent out into the tallgrass on the refuge. The task was to sweep the net quickly back and forth through the grasses to capture as many insects of as many species as possible. These had to be transferred for safekeeping into gallon-size bags. Every day, collections were made from nine or ten in the morning until about four in the afternoon. One mother and her ten to twelve chicks can eat about twelve bags of insects per day for the first few weeks of the chicks' lives. That is about one hundred insects each per day!

One Chick, One Victory at a Time

When I talked with Terry on the phone, he told me that, after all that hard work, only eighteen chicks survived. In fact, he said, they had thought the number was even lower, but then, in September, "four unbanded, un-radio-collared birds were seen." Obviously they, too, were survivors of the breeding season. It still seems a low survival rate—but it is eighteen more birds to boost the breeding colony.

I asked him what keeps him going, how he gets over the disappointments and setbacks they face in this quest to save the Attwater's prairie chicken. "Some days are more difficult than others," he said. And at those low moments, he thinks back on the "little victories" that they have experienced and so is able to regain his positive attitude. He praised the many volunteers who work so hard on behalf of this colorful and comical prairie grouse. "There is hope," he said, "so long as people are willing to help."

In Terry, the Attwater's prairie chicken has a powerful advocate. He has been directly involved with the birds since February 1993 and has no intention of giving up. His reasons for persisting? "I have always been drawn to the underdog," Terry told me, "and I like challenges. The Attwater's offers me both. And deep down, I want the Attwater's to still be around so my grandkids can enjoy them as much as I do."

Satiated vultures rest after feeding at the edge of Nepal's "Vulture Restaurant"—a place where vultures are given safe food, free of diclofenac. *(Manoj Gautan)*

Asian Vultures

Oriental White-Backed Vulture *(Gyps bengalensis)*
Long-Billed Vulture *(G. indicus)*
Slender-Billed Vulture *(G. tenuirostris)*

I have great respect for vultures. They fascinate me. I have not watched them in Asia, but I spent hours observing them on the Serengeti Plains of Tanzania. Their powerful flight is beautiful, their eyesight phenomenal, and their social behavior complex. The bare skin on the neck and head, which some people find repellent, is absolutely necessary—imagine getting blood and entrails clogging your feathers! And in some species, that bare skin acts as an indicator of mood. When an individual gets angry during competition at a carcass, or mating is involved, the neck may become bright pink! They are amazingly patient birds, too. Sometimes, having flown in from far away, they must watch while the bigger predators eat their fill—lions and then hyenas. Finally it is the turn of the vultures, which compete—often successfully—with the jackals.

Disaster Strikes in India

During the mid-1990s, Dr. Vibhu Prakash of the Bombay Natural History Society (BNHS) was one of the heroes who first alerted the scientific community to the fact that vultures in India were dying—mysteriously and in large numbers. Indeed, by the late 1990s, the three *Gyps* species—the long-billed or Indian vulture, the Oriental white-rumped or white-backed vulture, and the slender-billed vulture—were all listed as critically endangered. It was estimated that their populations had fallen by more than 97 percent in less than a decade—"one of the steepest declines experienced by any bird species," said Dr. Debbie Pain of the Royal Society for the Protection of Birds (RSPB). They were being found, dead and dying, in Nepal, in Pakistan, and throughout India. In some places, they had disappeared altogether.

Learning of the disaster, the Peregrine Fund sent scientists to monitor breeding populations of Oriental white-rumped vultures in Pakistan's Punjab Province. In 2000, they found twenty-four hundred occupied nests in thirteen breeding colonies. Returning to the same sites each breeding season, they recorded decreasing numbers occupying nest sites each year—and they were collecting dead vultures daily. By 2006, there were only twenty-seven breeding pairs. The report concluded: "This study has documented possibly the most disastrous population crash of any raptor species."

In 2007, when I was in India, I met Mike Pandey, a successful wildlife filmmaker and conservationist, and we talked about the vulture situation. He told me that when he first realized how endangered the Asian vultures had become, he decided to visit the carcass dump in Rajasthan where he had filmed vultures years before. Back then, he said, he had been literally engulfed by thousands of vultures, squabbling over the carcasses, cleaning the environment. But when he returned it was very different.

"I walked over the carcasses of thousands of vultures," he said. "I walked on the broken wings of the powerful birds." He was shocked. A pack of feral dogs that was feeding and breeding in the carcass grounds attacked him, but he managed to jump up onto the roof of his jeep, escaping with only a few scratches.

The Important Role of Scavengers

At one time, Mike told me, the Indian subcontinent had the highest density of vultures anywhere—close to eighty-seven million, he reckoned. At the same time, there were some nine hundred million cattle in India, the highest number in the world. Vultures used to clean up the carcasses of those that died in cities, villages, and the countryside—an estimated ten million a year. With fewer vultures, millions of cattle carcasses—and those of wild animals, too—now lie putrefying, creating a major health hazard for humans as well as livestock. The feral dogs and rats who took over the job of scavenging took much longer to strip a carcass.

Mike later sent me an e-mail noting that outbreaks of anthrax have recently been reported in four places across India. "Hot summer thermal currents could easily carry anthrax spores or pathogens from the decomposing carcasses into the stratosphere and carry it around the world," he wrote. Mike is genuinely fearful of what could happen if we lose the Asian vultures. "Our unthinking actions have knocked the master decomposer out of the skies," he told me. Without the vultures, "the putrefying carcasses are spawning grounds for hundreds of lethal mutating pathogens more dangerous than the bird flu or anything known to man."

Six months after my visit to India, I met with Jemima Parry-Jones, director of the International Centre for Birds of Prey in the UK. She commented that at the height of the vulture's decline in 1997, the World Health Organization estimated that thirty thousand people died of rabies in India—more than in any other country. And that, she said, could be attributable to the huge increase in rats and dogs, both rabies carriers. "It just goes to show that we have no idea how human-caused species declines will later impact humans."

One other service that vultures have traditionally performed in Asia involves their role in the funeral rites of some communities, including the Parsee of India. Jemima described an extraordinary meeting she had with a group of Parsees, including a high priest, in a rather noisy café in the UK. The Parsees explained how the decline in the vulture population posed a very real problem for their communities, as the vultures were relied upon to devour the bodies of their dead that, traditionally, are laid out in a circular raised structure known as the Tower of Silence. Gradually the chatter from the surrounding tables died away into a somewhat startled silence!

Why Were They Dying?

No wonder so many people were concerned about the possible extinction of the vultures in Asia—quite apart from their intrinsic value as a marvelously designed avian species. Initially it was thought that some disease was responsible, but postmortem examinations of dead birds failed to reveal any viral or bacterial infection. Affected vultures hunched their backs, their heads and necks drooped, and it was found that their internal organs were much inflamed, their livers covered with whitish crystals. It was assumed that the crystals were uric acid, and that the condition was similar to gout in humans. But what was causing it?

In May 2003, at a meeting of raptor biologists, a scientist working with the Peregrine Fund presented information that seemed to confirm a growing suspicion that the vulture deaths were linked to the anti-inflammatory painkilling drug diclofenac. Vultures that had died of gout had high levels of diclofenac in their kidneys. This medication, for veterinary use, had not been introduced to the Indian subcontinent until the early 1990s, but it had rapidly become very popular because it was cheap—less than a dollar for a course of treatment.

In January 2004, the results of a joint study conducted by the Peregrine Fund and the Ornithological Society of Pakistan confirmed that diclofenac was indeed the primary reason for vulture deaths. That was an important study that eventually resulted in a ban on the manufacturing of veterinary diclofenac by the Drug Controller General of India. This ban was soon introduced also in both Nepal and Pakistan.

Unfortunately, this is not enough: Not only are there major problems with enforcing the ban, but it is still legal to import, sell, and use diclofenac. Moreover, the diclofenac legally manufactured for human use has started to infiltrate the veterinary market. Until diclofenac has been completely removed from the environment in India, Pakistan, and Nepal, there is no safe future for the Asian vultures.

Nevertheless, the fact that the Indian government banned the manufacture of the drug, in such a relatively short time, was a historic triumph. In part, this can be attributed to the release, in March 2006, of a film made by Mike Pandey. Called *Broken Wings,* it resulted from his shocking visit to the carcass dumps. It is a powerful documentary that explains not only the reason for the vulture deaths but also the major role that these birds play in maintaining the health of the ecosystems of South Asia. It was shown, translated into five languages, on all the national TV channels. The radio carried the story as well.

At the same time there was personal outreach to the local people since, in the long term, they have the most influence over the vultures' fate. Mike told me that the Earth Matters Foundation created life-size vulture puppets and took them on road shows to rural communities, so that farmers and locals could see the magnificence of the birds and become sensitized to their plight. In parallel, the Peregrine Fund, the RSPB, and the BNHS produced and distributed more than ten thousand educational leaflets and flyers in Urdu and Hindi in villages closest to the remaining vulture colonies in Pakistan and India.

A Vulture Restaurant

Another initiative of the Peregrine Fund was to establish, in 2003, a "Vulture Restaurant" near a breeding site in Pakistan, where uncontaminated food was set out for the vultures. But although this reduced mortality in the peak breeding season, it made no difference once the young had fledged, and so it was closed. However, a similar vulture feeding station is still operated in Nepal by a dedicated group of Roots & Shoots members under the leadership of Manoj Gautam. The group is made up of local youth from Nawalparasi, a town about 150 miles west of Kathmandu. They gather animal carcasses (usually cows and buffalo) that are free of diclofenac and take them to their Vulture Restaurant to provide a supply of safe food for the birds. The work is hard—transporting the carcasses takes a lot of time, energy, and money.

Roots & Shoots is also working to raise awareness of the problem in local communities. As a result, Manoj told me, the people have become interested in helping to save the vultures. On one occasion in 2007, for example, some local youths reported to the Roots & Shoots group that they had found vultures eating an unidentified carcass. Manoj and his team immediately went to the spot and saw that more than half of the carcass was already eaten. Fearing that it might be diclofenac-infected, they buried what was left of it. Two days later they got the news that some vultures were sick and seemed to be dying. Again, the Roots & Shoots team rushed to the scene.

"We saw three vultures that were agonizing and flapping their wings on the ground and could not fly," Manoj said. One bird managed to fly away, but its wing beats were weak. The other two died. Diclofenac poisoning was confirmed when Manoj dissected the birds and found the telltale signs—uric acid in the liver and kidneys.

"With heavy hearts, seven of us buried the vultures in two pits dug by Roots & Shoots members in a nearby riverbank," Manoj told me. Fortunately, though, those deaths did not diminish, but rather strengthened, their determination. "We made a joint commitment," Manoj said, "that we will not let such destruction happen again." One major problem is that diclofenac is often smuggled across the border from India. And so, Manoj told me, the R&S members even patrol the local veterinary shops for diclofenac, doing their best to make sure no one is selling the drug.

The Threat of the Kite Festival

There is one other significant threat to the vultures—a very unexpected one. Once a year, throughout Asia, a series of incredibly popular kite festivals are held—a custom brought vividly to the Western world in

Khaled Hosseini's powerful best seller, now a film, *The Kite Runner*. These festivals are held in late winter to celebrate the harvest season. Kite-flying competitions are an old custom, but in recent times the traditional cotton thread has been replaced by strings coated with sharp powdered glass. Tens of thousands of kites darken the sky every day during the festival season. The kites compete, each trying to dislodge the other from the sky using the razor-sharp string to slice off opponents' kites . . . all in good fun.

Unfortunately, though, thousands of birds are hit and injured by the new kite strings, including many vultures. Mike Pandey told me the string called "Maajah" is the most dangerous—it sometimes shears off a bird's wing completely. He said that in just one day during the 2008 kite festival, more than eight thousand injured birds, including four badly injured vultures, were brought in by the local NGOs and volunteer groups in the city of Ahmedabad alone.

Even more tragic, Mike told me that these events take place at the very peak of the breeding season. "There is an urgent need to revert to the old cotton thread and also to ensure that no strings are left entangled in trees and bushes," he said. Fortunately, there is a ray of hope in this situation. Earth Matters Foundation, along with other concerned individuals and organizations, is fighting to get the glass-coated string banned all over the country. Also, Mike made a news feature about the vultures that was telecast on India's national network in spring 2008, appealing to people to stop using the Maajah string.

Captive Breeding: Is It the Solution?

During an international vulture conference in India in early 2004, a resolution was taken to start captive breeding programs for all three Asian species in order to save them from extinction. This was later ratified by the International Union for Conservation of Nature.

When I met Jemima, she told me that "in India we now have three facilities, one—the oldest—at Pinjore, outside Kalka in Haryana State; one in West Bengal; and one in Assam. The one in Assam will concentrate mainly on the slender-billed vulture as that is its natural range and it is the rarest of the three critically endangered species."

As with many birds of prey, obtaining eggs or chicks for breeding is seldom easy. "I will never forget when we were collecting chicks for the captive breeding program," Jemima told us. "We drove for miles down one of the scariest roads I have seen in a long time, and once we got to the nest one of the Indian villagers just took off his shoes, grabbed a hemp rope, and climbed an enormous tree to collect a vulture chick to be raised with one other chick in the program. I thought of my friends from the US who would want expensive ropes and carabiners to climb that tree."

In January 2007, the first white-rumped vulture chick hatched in Pinjore, but unfortunately it did not survive. When I spoke to Jemima in January 2008, she told me that more pairs of white-backed vultures were nesting in the facility, sitting on eggs. "These should be hatching soon," she said, "but it must be remembered that it takes time for the staff to gain the experience to get it all right."

There are currently 170 birds in the breeding program in India—about 40 in West Bengal, 4 in the new facility in Assam, and the rest in Pinjore. "We are aiming," Jemima told me, "to have seventy-five pairs—twenty-five of each species—in each of the facilities before we do any releasing, and of course the environment has to be 100 percent safe for them." Many of the birds are individuals that have been injured—especially during the kite festivals—and cannot be released again anyway.

Nepal is planning its own breeding facility, but not everyone is supportive. The pros and cons of captive breeding for ultimate release into the wild are, as we have seen, hotly debated in almost every case when a species faces extinction. Manoj is very excited by the recent attention and funding that have gone into protecting the Asian vultures, but believes that captive breeding should be a last resort, when there is only a small hope of saving a species in its natural habitat. And he believes the situation in Nepal is not so desperate that captive breeding is mandated. "We have recently observed positive signs about the vulture's situation," he wrote.

His main concern is that to start the breeding center, they plan to capture many birds; he is afraid that this would have a negative impact on the four hundred or so breeding pairs in Nepal. He is also skeptical as to whether captive-bred vultures will ever be able to learn the unique social and scavenging skills they will need in order to survive in the wild. "We need to conserve vultures as efficient scavengers, not as balls of flesh and bones covered in feathers that know nothing about scavenging," he told me. "They need to learn about their way of life, which is only possible if they are raised in the wild."

And so Manoj, and others who are against captive breeding in Nepal, would rather see conservation resources go into better protection of the breeding population in the wild, continuous monitoring of their nests, vigilance in detecting the sale of imported diclofenac, and fighting for legislation against the Maajah string in the kite festivals. All of this his Roots & Shoots team, assisted by other NGOs and increasingly concerned citizens, is already doing.

"Only When We Understand Will We Care"

If there is one strategy that almost all conservationists agree on, it is the role of education. Once people fully understand the vultures, realize the role they play in our lives, become sensitive to the glory of their flight, or simply fall in love with the charm of an individual, they are more likely to make real efforts to try to protect them. To this end, Manoj and the Roots & Shoots team are organizing the first "Vulture Watch Tour" in Nepal, from Kathmandu to the Vulture Restaurant in the Nawalparasi district. They hope this tour will raise funds for vulture conservation, and at the same time teach tourists about the bird's magnificence and the unique contribution it makes to maintaining ecological balance.

Mike Pandey, while making the film *Broken Wings*, learned to respect vultures as resilient and powerful scavengers and the supreme masters of the skies, and he, too, is dedicated to helping people understand these birds. "It is only when we understand something that we begin to respect it," he said. "And what we respect, we love . . . and what we love we protect and conserve." Education, he believes, is the key. People must understand "the dynamic law of nature and the fragile web that holds us all together in an interdependent cycle of life." He observed how, "when the people saw the link between their lives and the vulture, it changed them . . . reverence grew in the hearts of many, and they fell in love with a creature that was designed to keep the earth free from contamination, and free from disease."

Indeed, the vultures have eloquent and passionate ambassadors. It gives me hope that through captive breeding, better protection in the wild, and the increased vigilance and concern of the people, the Asian vultures will recover and once again, circling the air in their thousands, perform their ancient and crucial role in the great scheme of things.

Hawaiian Goose or Nene
(Branta sandvicensis)

The Hawaiian goose, or nene to give it its local name, is Hawaii's state bird. It got its name from the *ne ne* sound of its soft call. Scientists believe that it was once almost identical to the Canada goose, but after years of evolution the two species have diverged. The nene, with its long neck and black-and-cream markings, rarely swims. Its feet are only half webbed but have long toes suitable for climbing on the rocky lava flows of Hawaii. And since the nene evolved on a tropical island, with no need to escape either cold temperatures or predators, flying was less important for it than for the Canada goose—thus its wings are much weaker.

Prior to the "discovery" of the Hawaiian Islands by Captain James Cook, there were probably some twenty-five thousand or more nene. But during the 1940s, the species was almost completely wiped out by hunters, because there were no laws to prevent shooting the birds during the winter breeding season. In addition the usual invasive species, in the form of pigs, cats, mongooses, rats, and dogs, wreaked havoc as they preyed on eggs and young birds. The cats even killed adult geese. It is the same story for many large birds of the islands—without the ability to fly fast or far, they were easy prey for invaders.

By 1949, only thirty individuals remained in the wild. There were, however, other nene in captivity—some at the state endangered species facility at Pohakuloa, Hawaii, and some that had been sent to Slimbridge in the UK. Captive breeding began in these two sites for eventual return to the wild.

Recently I had a long talk with Kathleen Misajon, who has been working with the nene since 1995. After finishing her degree, she applied for

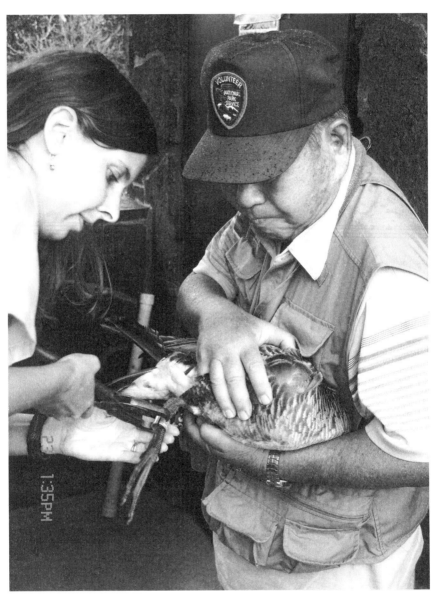

Park employee Kathleen Misajon with long-term (over twenty years) volunteer Lloyd Yoshina banding a wild nene at Hawaii Volcanoes National Park, 2006. (Ron McDow)

a three-month internship in Hawaii to continue working with the nene—and she is still there! Breeding the nene is not difficult, she told me, and since 1960 more than twenty-seven hundred have been raised and released. The problem—as for the giant panda, and many other species—has been trying to create a sufficient suitable and safe environment for their survival when returned to the wild.

Much of Hawaii's low-lying coastal areas have been developed, and that which remains is under continuous threat from further disturbances by humans and by invasive non-indigenous plants. But, says Kathleen, "Perhaps a bigger problem is that so much habitat was destroyed so long ago that no one really knows the exact components of ideal nene habitat." Maybe before all the human-caused disturbance to that unknown ideal habitat, the geese were better able to withstand periods of drought or heavy rainfall that are detrimental to them today, particularly during the breeding season.

The nene face many other threats, too. In addition to the ongoing problem of the introduced predators, increasing numbers of nene are being hit by cars. Unfortunately a major state highway cuts right through the park, and it separates an important nene breeding and roosting area from their feeding grounds. Normally the adults fly across, but when they have goslings they must walk, exposing themselves and their young to danger. It is the same when they are attracted to the grassy shoulders of roads after they have been mown. And those that venture onto the golf courses may even be killed by golf balls.

Kathleen told me she thinks the nene may never be 100 percent self-sustaining—the threats are too great. "However," she said, "the overall population is on the rise, and with proper management we can help sustain the wild populations."

Protection from Predators: "We Can't Just Abandon Them to Their Fate"

In the 1970s, a reintroduction program began in the Hawaii Volcanoes National Park. The areas selected for release were low-lying sites thought to be historical nene habitat. It was a very simple program: Several breeding pairs were kept in captivity, and when their young had fledged they were simply allowed to go free. Then, in the 1980s, additional young birds were released from the state breeding program into the park. During this twenty-year period, however, things did not go well for the youngsters when they were let out into the big wide world. This was not surprising, for the park attempted predator control only in the area immediately surrounding the release pens.

Thus the birds released into the wild saw high mortality and low breeding success. Clearly it did not make sense to go on breeding more and more young birds and abandoning them to their fate. A new strategy was developed that called for intensified predator control over a much larger area around selected breeding grounds. The next step was to erect fencing around one large nesting area and a suitable pasture to keep out the feral pigs that were suspected of killing many young birds as well as taking eggs, since the goslings were disappearing even when supplementary food was offered. Once four hundred acres was completely surrounded by a pig-proof fence, things improved, and during subsequent breeding seasons most goslings fledged.

Since the early 1990s, the population has grown to about two thousand individuals living in the wild, with the number rising each breeding season. They are living on four islands—Kauai, Maui, Molokai, and Hawaii. The nene is doing best on Kauai where there is no established mongoose population and grassy, lowland habitat is more available. Although small-scale captive releases still occur on Maui and Molokai, the current strategy focuses on minimizing the threats to the wild populations.

Now, Kathleen told me, they are experimenting with ways to keep out cats and mongooses using new fencing techniques. The design comes from Australia where so much work has been done on controlling predators of all kinds. The two-yard-high fence is constructed so that when a cat or mongoose climbs up from the outside, the wire curves outward and downward, leaving the marauder virtually upside down clinging to floppy wire netting.

Kathleen gave me an example of the danger posed by cats. It took place the day after Christmas 2001. She noticed a goose flying over an open lava field toward some vegetation where she felt she might be able to find its nest. As they are often remote and therefore not easy to find, she was excited as she trekked across the bare lava. Presently she came upon the gander, guarding his nest site. She moved on—and there she found the partially eaten female beside her eggs, now cold. The cat was still there, lying beside the carcass, glutted on goose meat. That was not the only proof she had of the hunting success of cats.

Nevertheless, the scientists and volunteers do not plan to give up. A few days after talking with Kathleen, I spoke with Darcy Hu, who has been working with nene in and around Hawaii Volcanoes National Park for more than fifteen years. I wanted a story with a happy ending—and she came up with one. It began on the day when she and her volunteer crew got a thirdhand report of a dog attack at Devastation, an area on the summit of Kilauea. They knew there were several nene there, including at least one banded pair with three partly grown goslings. The report merely indicated that the attack had involved at least one adult and one youngster.

Quickly they drove to the scene, but at first found no sign of bird or dog. Then they spotted and caught two goslings, too young to fly. At the same time, the call of an adult sounded deep in the forest. Not liking to abandon the two goslings—even if the calls were from a parent, there was no assurance the family would meet up, and the youngsters surely could not have survived on their own—they waited awhile. Soon the calls stopped. Although they searched for a while, they found no nene, and heard no more calls.

Still hoping the parents might show up, Darcy rigged a wire-mesh pen close to where the goslings had been found and left them there for several days, spying from a distance in case the parents returned. But there was no sign of any adult, and because the goslings were getting thin, they were moved to a captive facility. There, fortunately, an older nene couple was persuaded to adopt them. "Nene don't need help feeding themselves," wrote Darcy, "but they do have an almost physical need to be with other nene—you rarely see even unpaired individuals alone, and pairs and families almost always travel as a unit."

A few months after capturing the two young goslings, Darcy and her team spotted a pair of adult nene and a gosling about a mile from Devastation. Quickly they caught and banded the gosling and, as the parents stayed nearby, they were able to read their bands. "It was the missing parents and the third sibling of our two orphans!" wrote Darcy delightedly. The wild youngster was smaller and not as developed—food had surely been more plentiful and nutritious in captivity. But all the family had survived the dog attack. "We counted ourselves very lucky," Darcy wrote, "to have been able to conclude this particular story with a happy ending."

THANE'S FIELD NOTES

Cotton-Top Tamarin
(*Saguinus oedipus*)

Cotton-tops, at one pound, are among the smallest monkeys in the world. The first time I saw one was at the University of Wisconsin–Madison while visiting Dr. Charles Snowdon's cotton-top tamarin laboratory. I met a young grad student there named Anne Savage, who would eventually become the world's leading authority on this little monkey.

Nowadays, Anne often refers to cotton-top tamarins as little monkeys "with punk rock hairstyles." Working with them in captivity on a daily basis at the University of Wisconsin, she got to know them intimately and individually. Eventually she went to northwest Colombia to study their behavior in the wild for her PhD thesis research.

But of course, a squirrel-size monkey is hard to study from afar. And much like the squirrels in your backyard, they are extremely difficult to tell apart. So some of Anne's early research involved dyeing the white hair on the tops of the tamarins' heads so she could distinguish among them. This didn't hurt the monkeys; in fact, she used the same hair products used by people, just in much smaller quantities. And it is through these observations, as well as the use of innovative teeny little backpack radio transmitters, that Anne and her team unlock the behavioral biology of this endangered primate.

When asked what some of her favorite memories of her two decades observing cotton-tops have been, she chuckled and said, "Nothing is cuter than looking up in the trees after the babies are born. Cotton-tops almost always give birth to twins, and they are about the size of your pinkie finger with a long tail when they are born."

And Anne added that it's fun to watch them develop. Cotton-tops go through many of the same growth sequences that other primates, including people, go through. In fact, she said, "Babies go through a babbling time where they are practicing vocalizations all day long that eventually come to sound more like their parents'. They learn to use certain chirps or calls in the appropriate circumstances."

Today Anne and her team are trying to assess the population of cotton-top tamarins in Colombia. However, since they are still hunted for the pet trade, the monkeys run away from people, which means researchers can't simply walk through a forest and count the number of tamarins. So they use a trick learned from bird researchers and play vocalizations of other cotton-tops to draw them in. Unfortunately, the team has discovered that there are fewer tamarins than they had previously estimated. Anne told me that when they complete the forest surveys, it looks like there will be fewer than ten thousand tamarins remaining in the wild.

One of the reasons it's so important to protect the cotton-tops who still live in the wild is that they don't fare well in captive breeding programs. For some reason, they often develop colon cancer in captivity. Scientists are studying this and are still not sure why it occurs. It could be the stress of captivity or something missing from their diet that would ordinarily be found in the forest.

The good news is that when cotton-tops are given enough suitable habitats, they breed well on their own and can maintain a healthy population. "As a species, they don't tend to suffer high infant mortality," Anne told me. "So the secret is really in building on the reasons for local people to get involved in protecting the forest."

Which is why Anne founded Proyecto Tití, a remarkable group in northwest Colombia that works with the local community to protect the endangered cotton-top tamarin. The name comes from the Colombian word for "monkey," *tití*, and today the program involves dozens of Colombian biologists and students, as well as educators and community development efforts throughout the region.

Trouble with Plastic Grocery Bags

Early on in her fieldwork, Anne realized that the Colombian forests were shrinking due to a number of factors, including human encroachment. As communities move closer to the forests, they need to cut down more and more trees just to build their houses or have firewood for cooking their food. So one of the things that Proyecto Tití has done is come up with cheap and effective ways to help protect the forests, while also benefiting local people.

First, Anne and her team looked at how people used wood in their cooking. In most rural communities in Colombia, as around the world, they cook over an open fire. A family of five uses about fifteen logs every day to cook their meals. The Proyecto Tití team came up with a very simple cookstove called a binde, which is made out of clay. Instead of burning fifteen logs a day, they just need five logs to cook the same amount of food.

Another challenge of the local communities is that they have no way to manage their waste—in particular, the growth of plastic waste is inundating the area. Most noticeable are plastic bags—the kind you get at the grocery store—which are littered everywhere: roadsides, fields, and even in the tamarins' forest. But the bags weren't just an eyesore. They put wildlife at risk because animals come into contact with plastic that may have food items on it, or can transmit disease. Or sometimes animals even ingest the plastic bags, creating a nightmare.

So Proyecto Tití partnered with fifteen local women who are heads of households but did not have any consistent source of outside income. These women now crochet tote bags, not using wool yarn, but using plastic from the bags littering the ground. And while this sounds small-scale, these women have already recycled more than a million trash bags in the making of the "eco-mochilas," as they are called.

This solution-based program is a classic win–win, since the trash bags have become a valuable commodity. Anne pointed out that, "as the eco-mochilas grow in popularity, people from throughout the region know that they are helping to protect the tamarins and the forest as well."

Today there is a consortium of national and international conservation organizations working to protect the last remaining dry tropical forests in Colombia. Though the press most often references the dangers and drugs and crime in this South American nation, Anne pointed out that there is indeed hope for the future: "Most importantly, we are going to see a new protected reserve for cotton-top tamarins in coming years."

When I asked Anne what the future might look like for the cotton-tops fifty years from now in Colombia, she was optimistic. Not only have Proyecto Tití and other regional conservation groups helped to shift public pride and awareness, Anne said young people are taking a growing interest in conserving both wildlife and habitat. In fact, many Colombian students study wildlife biology in the United States or Europe and then return home to apply their knowledge. "What gives me hope," she said, "is to see that next generation really coming into fruition right now, developing long-term conservation plans to save species in Colombia."

THANE'S FIELD NOTES

Panamanian Golden Frog
(Atelopus zeteki)

If you have never held a common leopard frog, with its strikingly beautiful striped and shiny skin, you have missed one of life's great joys. Unfortunately, today, you would be lucky to *hear* a leopard frog calling, much less catch one.

There are many reasons for this, most of which people do not really understand. All around the globe, amphibians are under pressure—kind of like slimy canaries in the coal mine, warning us of hazards that we should heed before it is too late. Some blame climate change. Some blame UV exposure. But one thing for sure is that many amphibians are being killed by a chytrid fungus, *chytrid* being short for *Batrachochytrium dendrobatidis*, which attacks keratin in the dermal tissue of amphibians and suffocates them since they breathe through their skin. Scientists believe the fungus originated in Africa and was transported around the world in the 1930s by accident before anyone knew it even existed. It came on the backs of African frogs exported for medical research and the pet trade.

Infected frogs can be treated if you capture them and give them a special antifungal bath in captivity. Unfortunately, though, you can't treat the frogs and release them back into the wild, where the fungus is literally growing everywhere in some areas.

Perhaps the most dramatic amphibious rescue effort anywhere is one now famous in west-central Panama, where the very last of the golden frogs cling to life. The frogs, which have radiant orange-gold skin, have long been an important symbol of pride for Panamanians. The ancient indigenous people even considered them to be totems of prosperity and virility. Besides being valued for their folklore and beauty, the golden frogs happen to be important members of the region's ecosystem, as they primarily prey on mosquitoes and crop pests.

In an effort to protect this beautiful amphibian from extinction, a handful of sweaty and tireless conservationists set up a "frog Hilton," literally inside a hotel. The idea was to capture the endangered frogs in the nearby rain forest, cleanse them with the special bath, and then keep them in this quarantined hotel so they didn't die from the lethal fungus. What began as a very temporary rescue effort, eventually ended up taking up four rooms in the hotel and housing more than two hundred threatened frogs, along with the additional areas needed for food storage, volunteer staff, and expedition preparations.

This fascinating Hotel Campestre is also a favorite overnight destination for backpackers because of its immediate proximity to the forests and mountains at the edge of a dormant volcano's crater, about fifty miles southwest of Panama City. The two principal players in this unusual frog spa are Edgardo Griffith, a Panamanian biologist who has worked for years with endangered amphibians, and Heidi Ross, a Wisconsin native who first came to Central America as a Peace Corps volunteer. When they go searching, they often find more dead frogs than live ones, but they refuse to give up. After a year in the Campestre, the collection of frogs totaled more than two dozen species, all of them threatened by the fungus.

So this remote hotel became somewhat of a phenomenon for hikers and tourists as the legend grew that if you wanted to hear the raucous calls of male frogs, this was your last, best shot. Ross and Griffith ended up experts in amphibian husbandry—fixing filters and air pumps, as well as rearing tadpoles and various-size crickets and other insects to feed their brood. All the while, there was the nagging challenge of the long term. How would two people and a borrowed hotel make this work over the long haul? After all, Campestre couldn't house these frogs forever—and yet it wasn't safe to release them into the wild, where they would surely become infected.

Enter Bill Konstant and the Houston Zoo. Bill is the director of science and conservation for the zoo, and was able to rally support for the golden frog efforts. The support came in the form of volunteers and funding from numerous American zoos and botanical gardens, including the Buffalo Zoo, Cleveland Metroparks Zoo, and Rhode Island's Roger Williams Park Zoo. Amphibian experts not only joined the rescue mission but also helped to design the special facility that would hold the frogs and toads after their temporary stay at the Campestre. The new facility, called the El Valle Amphibian Conservation Center (EVACC), opened in 2007 and is located on the grounds of the El Nispero Zoo.

Bill is a combination rare in the field of wildlife conservation. Like many field biologists, he is highly educated and experienced, but he is also a scrapper and a doer. As he puts it, "Just because circumstances are dire for the golden frog and other amphibians, there is no reason to give up. In fact, it is time to raise the clarion call to action, because as long as there are frogs, there is hope." With a smile, he adds, "Besides, frogs know how to be frogs. That's their job. Ours is to figure a way to solve this mess so they can get back to their forests, streams, and wetlands."

Until it's safe for the golden frogs to return to the wild, the state-of-the-art facility will be the only safe haven for Panama's golden frog. In fact, organizers imagined the facility being a model for other threatened species that might need to be temporarily or permanently removed from the wild to be saved.

The question now remains—when will it be safe for the frog to return to the wild? Or will it ever? With persistence and gained knowledge, perhaps the streams of Panama will ring with the hopeful call of male frogs again. Time will tell.

PART 4

The Heroic Struggle to Save Our Island Birds

Introduction

Ever since that time long, long ago when humans first set off in flimsy boats to explore the Seven Seas, island species have been at risk. Many of these animals, insects, and plants evolved over millions of years, perfectly adapting to the environment in which they lived—an environment without competition from terrestrial predators or trampling herbivores. Some birds, like the well-known species on the Galápagos, never needed to develop flight-or-fight behaviors, never learned to fear.

And so, from the start, seafaring humans—whether they stayed and colonized an island, or merely paused in passing to stock up on water and food during long sea journeys—found island birds easy prey. The flightless dodo was eaten to extinction; the flightless kakapo nearly was.

Settlers brought their livestock with them, mainly goats and pigs. Rabbits were introduced to provide food and quickly multiplied. Stoats, imported to hunt the rabbits when their populations got out of hand, found island fauna easy prey. Cats initially provided pest control services when rats disembarked from visiting ships but soon, adapting to a feral lifestyle, began to hunt the unsuspecting birds. Many alien species of plants were introduced, some of which adapted quickly to the new conditions and spread. The native animals and plants simply could not cope with such unexpected invasions. The delicate balance of nature was interrupted, again and again, with disastrous results. Countless island species disappeared along with the dodo; countless others were brought to the brink of extinction.

While doing the research for this book, I met and spoke to some extraordinary and dedicated people who have been fighting to turn back the clock on these islands. I have been learning about the Herculean efforts required of them as they struggle to save unique and very precious life-forms, both animal and plant, from extinction. It cannot be done without hard work, absolute commitment, and a willingness to face hardship and sometimes danger. And one of the most difficult, challenging—and often controversial—aspects of their work is, of course, the task of removing alien species from island habitats.

In other words, these biologists have been forced over the years, and around the world, to poison, trap, or shoot thousands and thousands of innocent creatures. They must not relax. The work is intensive, and usually very expensive. The same techniques cannot be used on all. The larger ones, like goats and pigs, can be hunted. Cats can be shot initially, but as their numbers are reduced they have to be trapped. Rats are even more difficult, mainly because of their sheer numbers—only poisoning has so far been effective. And there is always the possibility, with both trapping and poisoning, that the wrong animals will be killed, especially native rodents. On one island in the Pacific, the bait was taken by land crabs—it did not harm them, but hundreds of rats escaped. On Canna Island in the Hebrides, biologists evacuated 150 endangered canna mice (a distinct subspecies) before successfully exterminating the approximately ten thousand brown rats that had invaded this small island. (The mice will soon be reintroduced.)

"Pest" Species Versus Endangered Species

It is not surprising that the large-scale eradication of so many luckless creatures has led to opposition from many of those concerned for animal rights. They argue, with justification, that the welfare of the "pest" animals is not adequately addressed. The biologists are accused of cruelty and indifference to the suffering of living beings who also have a right to exist. After all, none of them *chose* to invade the islands where, when given free range, they set about living off the land. Unfortunately, this was very destructive. Goats are particularly skilled in this respect. They are intelligent and adaptive. They need little water and can eat almost anything. When they have finished off all the ground foliage, they even climb trees. Rabbits, while smaller in size, are far superior in their ability to multiply. And think how even a well-fed domestic cat can inflict severe damage on local bird and rodent populations. On an island, the impact of feral cats can be devastating.

My friend Don Merton, who has been involved with restoring islands for decades, told me how, in the late nineteenth century, the lighthouse keeper's cat on Stephen's Island, New Zealand, killed all eighteen of the last Stephen's Island wrens known to science, and laid them at its owner's doorstep. This wren was just one of the countless endemic species exterminated by animals unwittingly taken to islands by humans.

But, let me repeat, none of the introduced species went voluntarily to the islands. They had no more choice than the early cargoes of prisoners off-loaded at Botany Bay. We put them there. Just as we put mongooses in the Virgin Islands to kill snakes. We put arctic foxes on the Aleutian Islands where, safe from their predators, they could breed and provide skins for the fur trade—while at the same time decimating some of the island fauna and damaging the whole ecosystem. We took European red foxes to Australia so that people could hunt them with horses and hounds—and the foxes hunted the smaller indigenous marsupials and birds. The only crime of these so-called pest species is that they have been—just like *Homo sapiens*—too successful.

It comes down to a conflict between concern for the individual and concern for the future of a species. Even the needs of individuals within the population being saved are sometimes subsumed for the good of the species. Animals raised in captivity may be released into the wild in the certain knowledge that 30 percent or more will not make it. I have always been an advocate for the individual. But after learning how some of the efforts to save the very last members of an amazing and unique species— such as the kakapo or the Zino's petrel—almost failed because of cat predation, and looked at the utter destruction caused by goats and rabbits, I had to rethink my position.

If only there were really humane ways of removing the alien species. But sterilization, as sometimes practiced with stray dogs and cats, simply wouldn't work, and even if you could live-trap all the predators— where would you put them? What could you do with shiploads of pigs and goats that had been corralled? If only the unfortunate invaders had never been introduced, if only there was an ethical way of removing them. But they were, and there isn't—and they have to go. After all, as Don said to me, alien predatory animals in order to survive must kill hundreds if not thousands of native birds and other wildlife each year— so causing suffering that is unseen and ongoing.

And even though I grieve for the slaughter of the invaders, I am filled with admiration for the persistence of those who work so hard to remove them from the islands. Don Merton, who first succeeded in eradicating rats from islands in the early 1960s, was a true pioneer in techniques for removing alien species. He developed methods for removing invasive species that have been modified for eradication projects around the world. No one wants to devote themselves to killing—yet as we have seen, to protect the birds and their defenseless young it must be done.

All of these island birds exist only because of the determination and ingenuity of those who refused to let them die. I have attempted to do justice to the extraordinary men and women who have saved these island birds from joining the dodo in the void from which there is no return. They have endured many setbacks. They must be patient, persistent, and resilient, as well as tough and courageous—and possibly a little crazy. And as you will see, they are.

Black Robin or Chatham Island Robin
(Petroica traversi)

My story of the black robin began when I met Don Merton in the early 1990s. He is quiet and soft-spoken and, like so many people who have accomplished extraordinary things, he is modest. Don had been invited to a reception held to welcome me to New Zealand, and we were not able to talk long. But he gave me a glimpse of the fascinating work he did, and his passion for saving endangered birds. The rest I have learned from subsequent chats on the telephone and e-mail correspondence. And of course, from reading about his work.

His love affair with wildlife began in the 1940s when he was a small child, growing up on the east coast of New Zealand's North Island. "From the age of about four years," Don told me, "I was nutty about wildlife and spent much time watching birds, lizards, and insects—especially looking for birds' nests." When he was five years old, his grandmother came to stay and brought with her a canary. "That little yellow bird sang its way through the 1940s and . . . ignited my passion for birds," Don said. One day he and his brothers "gave my grandmother's canary a (European) goldfinch chick to foster. It adopted the chick as its own and raised it." Thirty-five years later, his recollection of this incident eventually saved the black robin species from imminent extinction. (I shall tell more about that later.)

He was twelve years old when he made the decision to devote his life to trying to save birds that were in danger of extinction. And he certainly followed his dream, beginning his career in 1960 (the same year that I arrived at Gombe National Park in Tanzania) and playing a key role in the rescue and recovery of some of his country's—and the world's—most

Don Merton with one of his beloved black robins. A childhood memory of his grandmother's canary helped Don figure out a way to save this endearing bird from imminent extinction. *(Rob Chappel)*

endangered birds. It all began in 1961 when he spent a month on Big South Cape Island—now known by its indigenous name Taukihepa (off the southwest coast of New Zealand's Stewart Island)—which still retained its full quota of indigenous wildlife. Indeed, along with two other tiny adjacent islands, it was the final refuge for several animals formerly abundant and widespread on the mainland, including the South Island saddleback.

Rats and Other Invasive Animals

That trip, along with subsequent field trips to remote areas, led Don to wonder why it was that on mainland New Zealand, despite hundreds of thousands of acres of seemingly intact forest and other habitats, native wildlife was in such a predicament. Why had massive extinctions and reductions in the range of so many species occurred? Don and some colleagues were convinced that the impact of predatory mammals introduced by European settlers, on purpose (such as cats, ferrets, and stoats) or accidentally (rats and mice), was the primary reason. But some leading biologists (educated in Europe or North America) argued strongly that predation was natural, and it was habitat loss that was primarily affecting wildlife in New Zealand.

Gone Forever

Then something happened that, in Don's words, "not only clinched the argument, but changed forever the way we were to perceive, protect, and manage our islands and their native plants and animals." In March 1964, three years after Don visited Taukihepa, he heard that ship rats had reached the island and increased to plague proportions, causing the wildlife to suffer massive damage. Don and his colleagues, anticipating "a biological disaster," wanted to do something about it, but some of the most respected biologists refused to believe that the rats posed a significant threat to wildlife, and vigorously opposed any suggestion to intervene. They argued that any intervention would "change the ecology in a way that we cannot predict: We should intervene only after research has shown there is in fact a problem."

Eventually, after five months of arguing, and thanks to the support of some senior Wildlife Service staff, Don and his colleagues were given permission to set off on a rescue mission. "We were successful in saving the saddleback through transferring some of the remnants to two small neighboring pest-free islands," reported Don. But they arrived too late to save the bush wren, the Stewart Island bush snipe, and the greater short-tailed bat, along with an unknown number of invertebrate species. They were gone. Forever. However, the saddleback now numbers in the low thousands and flourishes on more than a dozen islands. It was the first bird species to be rescued from imminent extinction and restored to viability in the wild through direct human intervention.

"The tragedy of Taukihepa was a valuable and timely lesson for this, and other aspiring conservation workers," Don wrote to me, "and served to convince even the most skeptical that, unaided, rats are capable of inducing ecological collapse and extinction within native island faunas." Indeed, that disaster led to the development of island quarantine protocols and methods of predator eradication and control that have made it possible to maintain biologically important islands free of pests.

Over the years, Don has helped to save many birds from extinction. One drama still ongoing, in which Don has played a major role for many years, is the fight to save the kakapo, the only flightless parrot in the world. It is absolutely fascinating and is described in full on our Web site (janegoodallhopeforanimals.com). Don also played a key role in the rescue and recovery of the Australian noisy scrub-bird, the Seychelles magpie robin, and other animals endemic to the Seychelles Islands of the Indian Ocean.

An Incredible Story

Of all Don's accomplishments, the saving of the black robin is the one I love best. "Black robins," said Don, "are delightful, friendly little birds that have an affinity with people—often approaching to within a meter, even perching briefly on one's foot or head! They quickly capture the heart of even the most unenthusiastic bird observer! I just loved them and, as well as feeling very privileged, felt a massive responsibility to current and future generations around the world, to save this fantastic little life-form from the brink of extinction."

What a tough job that turned out to be. Since the 1880s, black robins were confined to Little Mangere Island, a tiny rock stack in midocean off the Chatham Islands, about five hundred miles east of New Zealand. Here, in their last refuge, they lived in just twelve acres of woody vegetation. It was thought that they were safe, at least in the short term,

until in 1972 a team of biologists captured and color-banded every individual—and found that there were only eighteen in all. In subsequent
years numbers continued to decline, and Don advocated immediate intervention. "But I was overruled," he told me. Some thought the downward trend was part of a cycle, and that numbers would soon recover
unaided. Only in 1976, "when there were just nine black robins left in
the world, was there general agreement that action should be taken."

Don told me that he and most of his colleagues "felt very strongly
about what should be done, and often there was frustration at not being
permitted to get on with it." When, finally, they got the go-ahead to
capture and relocate the remaining robins, they reached the island in
September 1976 to find just seven birds left—and only two of them
were female. And only one of the females would prove to be productive.
This female, marked with a blue leg band, would become famous as Old
Blue. The tiny group of survivors was moved from Little Mangere Island, where their scrub forest environment was dying and no longer able
to support them, to nearby Mangere Island. This was but the first step
in a dramatic and ultimately successful attempt to rescue the species.

Old Blue—The Matriarch Who Saved Her Species

Black robins normally mate for life. Old Blue and her mate nested during the next breeding season, but their eggs were infertile. Amazingly,
Old Blue then abandoned her longtime partner and in his place selected
a younger male soon to be known as Old Yellow (because of his yellow
leg band). Again Old Blue laid eggs—and now this little family became
part of Don's innovative cross-fostering program.

It was that childhood memory of the canary fostering the goldfinch
that gave Don an idea for how he might be able to boost the normally
low productivity of the species. In normal circumstances, a black robin
pair rears no more than one brood of two chicks per year, so the species
lacks the ability to recover quickly from adversity. But if a nest was destroyed, or eggs taken, the black robins would build a new nest and
produce another clutch. So Don destroyed the nest, removed both of
Old Blue's eggs, and placed them in tomtit nests, where they were successfully fostered.

Old Blue and Old Yellow then made a second nest, and she laid a second clutch. Again the eggs were taken. Meanwhile, the chicks from her
first tomtit-hatched clutch were returned to Old Blue so they would learn

behavior appropriate to their species. Then the second clutch hatched. Don told me that when he returned them to join the first lot, Old Blue looked up at him with a resigned expression, as if to say "Goodness, what next?" Whereupon he reassured her, "We shall help you feed them, love, don't worry." I have always cherished the mental picture of Don and his team rushing around searching for suitable food for the artificially extended family of black robin chicks they had helped to create.

The same procedure was repeated for the next few seasons, thus giving the single family group of black robins a kick start. "Cross-fostering proved highly effective," Don said, "but at the start the technique was untested and thus of high risk. . . . If we failed, we would be blamed for exterminating the species!"

Desperately Don and his team worked to save these birds. "Old Blue, Old Yellow, and their many chicks became my extended family," said Don. "I thought about them constantly. While in the field—often for months at a stretch—we spoke about little else." Each spring, when Don visited Mangere Island, he couldn't wait to find out which birds had survived the winter. "Each new nest, egg laid, or chick hatched was cause for celebration, and any deaths were almost the equivalent of a loss within the family!" He never enjoyed the times when, to ensure their long-term survival, he had to take their eggs and destroy their nests.

Old Blue finally passed away in 1984. She lived to be thirteen years old, more than twice the life span of most robins—despite the abnormal number of eggs and chicks she had been manipulated into producing. And because her story had touched the hearts of many New Zealanders, a plaque was set up in her memory at the Chatham Island airport, and the Right Honorable Peter Tapsell, minister for internal affairs, announced the death of "Old Blue—matriarch & savior of the Black Robin species." National and international media broadcast the story of the world's rarest and most endangered bird who had in her "geriatric years" brought her species back from the brink.

A Bright Future

By the late 1980s, numbers of black robins had increased beyond the one hundred mark. Groups of black robins were then established on an additional island. After this, there was no further need for intensive, hands-on management of the birds. Don told me there are now approximately two hundred black robins on two islands. All are descended from just one pair—Old Blue and her mate Old Yellow—thus in their genetic profiles all are as identical as identical twins.

"Thankfully," said Don, "there are no apparent genetic problems." However, habitats on the two islands are at saturation point, which means that the species cannot increase in number or expand in range. Also, during and after each breeding season, there is considerable wastage—young birds die because they have nowhere to live. Don has long advocated re-establishment of a population on Little Mangere Island—the very place whence he removed the last members of the species at the start of the rescue. Since then Little Mangere's woody vegetation has recovered, and being free of predatory mammals the island presents—in the short term at least—the only available option for black robins in the Chatham Islands. Don strongly supports this proposal. "And needless to say," he told me, "I would love to be involved!"

Christmas Island Park Manager Max Orchard and his wife, Beverly, have devoted the past sixteen years (and even handed over their yard and carport) to nurturing injured or orphaned Abbott's boobies. Here Max is feeding fish to a recovering juvenile. *(Corey Piper)*

Abbott's Booby
(Papasula abbotti)

The Abbott's booby is an ancient species, a true oceanic bird, living at sea and coming ashore only to breed. It nests only on Christmas Island (a territory of Australia), a fifty-million-year-old extinct volcano rising out of the Indian Ocean, ten degrees south of the equator. Abbott's boobies are impressive-looking birds, with bright white heads and necks, long dark-tipped bills, and narrow black wings. Growing as large as thirty-one inches in length, they are the largest of the boobies—some call them the "jumbo jet" of the boobies.

These boobies have a life span of up to forty years, and the young birds do not start to breed until they are about eight years old. They have one of the longest breeding cycles of any bird (fifteen months), so breeding occurs at two-year intervals. They nest in the tops of trees, laying just one egg.

Their numbers began to decline when, in the 1960s, phosphate mining began in full force on Christmas Island. In order to mine the mineral, it was necessary to clear large strips of the primary forest—interfering with the boobies' breeding, since they nest in the tops of forest trees. These tall trees often grew over the richest phosphate deposits, so that Abbott's boobies were in direct conflict with mining interests. The boobies have thus lost the greater part of their historic breeding habitat. Their population is now estimated at about twenty-five hundred breeding pairs.

Although the local government as well as the mining company tried to monitor and protect the habitat and nests, the Abbott's booby continued to decline. Finally in 1977, Don Merton, well established as an island restoration expert by then, was sent to Christmas Island to advise the Australian government and the British Phosphate Commission on wildlife conservation matters. He spent two years with his young family on Christmas Island and ultimately helped convince the government to create the island's first biological reserve, a four-thousand-acre national park built in 1980—one of the largest and least modified raised tropical island rain forest ecosystems to be protected anywhere. Another conservation initiative on Christmas Island was the plan for a comprehensive program monitoring the breeding and conservation of Abbott's booby.

Destroyed Habitat and Chicks in Peril

By the mid-1980s, it was estimated that some 33 percent of the habitat formerly used by the boobies had already been destroyed, and mining activities had created at least seventy clearings in the forests. Not only had this deprived the boobies of nest sites, but it was found that birds nesting near the clearings suffered from wind turbulence. Sadly, this caused unfledged Abbott's booby chicks to be blown from their nesting sites. Strong winds can sometimes blow fledgling and even adult boobies from branches, and if a bird falls to the forest floor it will die unless it manages to climb up through the vegetation. These birds can take off from the ground, but with great difficulty. They need sufficient wind from the right direction and a clear "runway" to get airborne. Unless found and rescued, they are normally doomed.

Ultimately, it was decided that the best way to protect the boobies was to protect and expand the island forests, by returning precious topsoil and replanting areas cleared for mining. Hopefully this would reduce the wind turbulence that is so detrimental to nesting boobies. Thousands of seedlings were raised and planted, using funds from the mining companies negotiated as part of their agreements.

The Restoration Program Comes Under Attack

Shockingly, three years later, the area given top priority by wildlife biologists was selected by the government for an immigration reception and processing center. Not only that, but the section of the mine site that had already been reforested was cut down. This has sparked a great deal of anger in the conservation community, particularly among those who have worked so hard on this restoration program.

The National Parks Australia Council has denounced the plan as "illegal" and requested that work on the site should cease immediately since it did not have proper approvals. "There are more suitable sites on the island that do not have such severe environmental impacts, and already have infrastructure provided," said Andrew Cox, president of the council.

And Monash University biologist Peter Green, one of those originally involved in the Abbott's booby monitoring program and with a long association with the island, commented that "the Abbott's booby birds were the focus of a commonwealth-funded rehabilitation program, which had been taking place at the site of the new detention center. And now," he concluded, "they have just put a bulldozer through it."

Not only this, but the government is actually negotiating new deals with the mining company. In 1988, the federal government had ruled that there would be no further clearing of rain forest on Christmas Island; the company is now appealing that ruling, and has recently sought permission to expand its lease to include new areas of old-growth forest. "It's crazy," said Andrew Cox. "Christmas Island is a jewel in the environmental crown of Australia [with] the world's only population of Abbott's booby birds and other endemic creatures . . . and we should protect it." It's one of very few raised tropical island ecosystems remaining anywhere.

For now, the Abbott's booby numbers seem secure. But this latest environmental blow could prove harmful.

The Orchard Nursing Home and Orphanage

Meanwhile, for the past sixteen years, amid all this Christmas Island turbulence, Max and Beverly Orchard have been rescuing the island's injured and orphaned endangered birds. Max has been a wildlife ranger for more than thirty years, working initially in Tasmania. He and Beverly have spent most of their adult lives rescuing and caring for orphaned or injured animals, with a special interest in endangered species. When they were in Tasmania, they used to care for wombats, wallabies, and Tasmanian devils.

I have talked with them on the phone, and the warmth and passion of their caring personalities reaches me all the way from Christmas Island. Beverly explained that every time a big storm hits the island during nesting season, many of the young ones fall out of their nests. It's during the monsoon season that there are so many casualties—that's March through

Beverly Orchard is the "heart and soul" of the operation, according to her husband, Max. "She can get along with the fiercest of them." *(Max Orchard)*

August. But the injured and orphaned keep coming until Christmas. Visitors to the park and local hikers find the birds and are always guided to Max and Beverly. Nestlings grow exceedingly slowly, remaining in their nests for about a year, so they are vulnerable for a very long period.

When they arrive, "they are often dehydrated, starving and completely depleted—but they can be resilient," said Beverly. The Orchards take the little ones and those that are injured into their home and put them into small nesting boxes. Then Beverly nurses them, giving them water and small fish from the huge stock that they keep in the freezer. She soaks the fish extra-long in water so they're easier for the young ones to swallow. If birds are injured, Max will try to heal them—say, trying to repair a broken leg. One time he managed to surgically remove a fishhook from an Abbott's booby gut.

Of course, inevitably, a number of her patients die. But Beverly is amazed by the boobies' resiliency. "We've had a number of them come in that I didn't think had any chance of making it," she said. "Some couldn't even lift up their heads." When she left them for the night, she'd felt "sure they were breathing their last breaths." But after her nursing and a night's rest, she'd check on them in the morning "and they'd be peering out at me, talking excitedly—hungry for breakfast."

Nesting in Plastic Chairs

Each patient has its own nest—an "old plastic office chair" kept outside under Max and Beverly's carport. They realized that this was the most comfortable spot, especially as feeding time can be very messy. At any

Some of Max and Beverly's family of juvenile boobies waiting for breakfast. *(Bev Orchard)*

Young patient recuperating on its office chair nest in the Orchards' carport. *(Dr. Janos Hennicke)*

given time, there are dozens of plastic chair nests lined up out there. After an injured or orphaned booby has been nursed back to health or come to a certain stage of maturity in its box inside the house, Beverly and Max try to transition it to a plastic chair as soon as possible.

In the wild, boobies nest in extremely high trees. "We try to replicate what happens in the wild," said Max, "but there's no way we can replicate the nest. We figured out that the best plan was to give them each a plastic chair, and we feed them fish and squid—the same kind of food we believe their parents would feed them in the wild." They fly out of their chair nest every day for a few hours, always coming back for feeding times.

"They are usually quite friendly and cooperative birds," said Max, "but woe be it to any booby who sits on the wrong chair nest!"

There Are Boobies and Boobies

"They all get the same name—Eric," said Max. This is based on the Monty Python skit "Fish License," in which John Cleese plays a man who names all his pets Eric. But the boobies definitely differ from one another.

"Each one has its own personality," said Beverly. "Some of them like to be held and are quite smoochy. They are very conversational birds and like to talk with their parents, so when it's feeding time, I always go out and talk to them. 'How are you?' 'How was your day?'—that kind of thing. They all start squawking back—they all get very excited to talk with me." They have a croaking-bellow sound that Max jokingly noted sounds like someone getting sick—"kind of a retching sound."

"We try not to handle them too much," said Beverly. "Once the babies get their feathers, we put them out on a chair and don't handle them any longer. This way when they leave us, they won't be tempted to land on boats and visit with other humans."

Max calls Beverly the "heart and soul" of the operation. "She can get along with the fiercest of them—the ones who come in screeching and strutting menacingly," said Max. "Before long, she has them all calmed down and practically cooing when they see her."

Over the years, this amazing couple have rescued close to five hundred Abbott's boobies in all. They mature slowly—it's about a year until maturity—and those the Orchards deal with are usually in recovery, so their development is even slower. Some stay nested on their plastic chairs with Max and Beverly for up to two years. And then, finally, they are ready for life in the wild.

"The day comes when they are finally mature and they take off, and that's the last you'll see them," said Beverly. Fortunately, though, before they are ready to go the boobies have a good-bye ritual so Max and Beverly can prepare for the departure: "One day they will come back to the chair, but not eat," said Beverly. "And they will suddenly be especially talkative—as if they have a lot to say. This is when we know they have found a food source—they are finally self-reliant. Perhaps they are telling us about what they've found or thanking us or just saying good-bye. We have no way of knowing" she added. "Then they'll sleep peacefully through the night on the nest, say a final good-bye in the morning, and take off for good."

"They become part of our family," said Max. "They're completely dependent on you and then they go off forever. It's a mixed feeling. You're happy that another one is returned to the wild—this is why we do all this work. So of course you hope it all goes well for them, but it's hard never seeing them again after they've been a part of your family for so long."

Max told me that apart from their habitat problems, the latest threat to the boobies is the high number of nearby fishing operations that are depleting their food resources as well as posing a direct threat through nets and long-line fishhooks. The Abbott's booby may be saved from extinction for now, said Max, "but we need to remain vigilant."

This adult cahow climbed on biologist Jeremy Madeiros's head before taking off. Jeremy's head was the best perch this cahow could find in the treeless habitat of Castle Harbor, Bermuda. *(Andrew Dobson)*

Bermuda Petrel or Cahow
(*Pterodroma cahow*)

I have been fascinated by petrels ever since I read *Tom the Water Baby* as a child. In that old classic, it was the Stormy Petrel, called "Mother Carey's Chicken," who came into the story. Mother Carey is the name petrels have long been called by sailors, who meet them far from shore, at home in the wilderness of the oceans. The name is thought to be derived from *Mater Cara*, which is how the early Spanish and Portuguese sailors, the first Westerners to sail the southern seas, referred to the Virgin Mary. And *petrel* is thought to refer to Saint Peter, because when they feed, the birds seem to be walking on water.

The subtropical petrel whose story I share here, the Bermuda petrel, is one of the so-called gadfly petrels belonging to the genus *Pterodroma*— from the Greek *pteron*, meaning "wing," and *dromos*, meaning "running": hence "the winged runner." This recognizes the fast, acrobatic, and gliding flight. Indeed, all petrels are masters of the air, able to survive fierce storms and fly through howling winds with the wild spray of giant waves crashing below them. It is when they come on land to breed that they suffer so terribly from the damage that we have inflicted on their island environments.

The local name for the Bermuda petrel is the cahow, a word said to derive from the species' eerie nocturnal cries. These initially pro- tected Bermuda and its isles from settlement, because the Spanish sailors believed they were inhabited by evil spirits. Indeed, Bermuda was once referred to as the "Isle of Devils." In those days, in the early 1500s, when Bermuda was discovered by the Spanish, it is estimated that at least half a million cahow returned each breeding season to the coastal forests of Bermuda and the surrounding islands, nesting in burrows in the sandy soils.

Unfortunately the "evil spirits" did not prevent the sailors from landing in search of fresh food and water. And they put pigs ashore to breed to provide a future supply of fresh meat. Thus began the destruction of the cahow's nesting grounds. And then things got worse. The British, quickly realizing that birds rather than spirits produced the strange sounds, began to colonize the beautiful tropical island, and the early settlers brought with them the usual invasive species. Also, year after year, while the petrels were away at sea, the British took over the cahow's nesting grounds for farming. And when the petrels returned for the breeding season, they were killed for food in great numbers—despite the birds being given official protection when the governor made a proclamation "against the spoyle and havocke of the Cohowes." Surely one of the earliest conservation efforts ever!

By 1620, it was believed that the cahow was extinct. Except that, just occasionally, someone reported otherwise: In 1906, for example, a cahow was actually collected, although it was not identified as such at the time. And then in 1935, a fledgling cahow hit a lighthouse—and its dead body proved conclusively that, somewhere, a population still survived. World War II temporarily put an end to speculations about their existence. But the death of that fledgling caught the imagination of a local schoolboy, David Wingate.

The Cahow Lives

"The year that fledgling collided with the lighthouse," David would remember, "was the year that I was born." He told me this during a telephone conversation in 2008. He vividly recalled sitting in his kayak one day, looking toward the islets beyond the lighthouse, and thinking: "It was only fifteen years ago when that young cahow died. Perhaps, just perhaps, they are still out there. Somewhere." He told me that the hairs on his neck stood on end at the thought.

Nor was he alone. Dr. Robert Cushman Murphy, of the American Museum of Natural History, managed to get funds for a thorough survey to find out, once and for all, the truth about the cahow. And when, in 1951, he set out along with the director of the Bermuda Aquarium, David was invited to join them. What a thrilling day for a sixteen-year-old schoolboy to be present when they came upon seven nesting pairs of Bermuda petrels on a tiny islet off the Bermuda coast. (Subsequently they found eleven more pairs on another three islets.) "I could hardly believe in my good luck," David said. "It was a dream come true. And from that moment I knew my life's path."

Somehow the cahow had survived against seemingly impossible odds—but there were so few of them. Could the newly rediscovered colony possibly survive much longer? Without the determination and energy of David Wingate, who devoted much of the rest of his life to their cause, they probably would not have, for their situation back then was desperate.

The four tiny rocky islets (off Castle Harbor, east of Bermuda) where the tiny remnant of the once huge cahow population had been forced to nest shared a total area of only just over two acres. Moreover, these islets were, to all intents and purposes, devoid of vegetation, and the small, shallow pockets of soil were quite unsuitable for nesting burrows. The cahows were laying their single eggs and raising their single chicks in rock cavities almost at sea level. And the islets, situated at the edge of the protective reef, were subject to severe battering by stormy seas. On top of all this, during the 1960s high levels of DDT were measured in both chicks and eggs, and this almost certainly had an adverse effect on their reproductive success. Indeed, according to David it reduced breeding success by about half. (David had become involved in the fight to ban DDT described in part 2 in the chapter on the peregrine falcon.)

And finally, as if all this were not enough, the petrels suffered in competition with the larger, more aggressive, and still common white-tailed tropicbirds. The cahow lays in January; chicks hatch in March. The competing tropicbird nests later, and on finding a nest site occupied will force a cahow chick out and take over. In some years, petrel chick mortality has been as high as 60 percent as a direct result of this competition for nest sites on the inhospitable islets.

A Nesting Real Estate Business

One of the first steps taken to assist the few remaining cahow was to fit each existing nest site with a wooden baffle that prevented the entry of the larger tropicbirds. Next, a number of artificial nest sites were constructed, each consisting of a long tunnel ending in a concrete chamber. Both these measures led to increased breeding success. And from that time on, the biologists working to save the cahow have ensured that there are at least ten extra nests ready for each breeding season. It has been necessary also to repair those damaged by the storms that have become worse due to rising sea levels. "Before 1989," David told me, "we never had real problems with flooding." But in 1995, some 40 percent of nests were damaged by a hurricane; in 2003, when the area was devastated by Hurricane Fabian, 60 percent of nest sites were destroyed and massive chunks of the islands were lost. It was fortunate that the hurricanes occurred when the cahow were at sea.

Because of the worsening situation, a new set of nest burrows was constructed on the most elevated part of the largest nesting islet—eight feet higher than the nests destroyed by Hurricane Fabian. Breeding pairs found scratching in the debris of their old nest sites were attracted to the new site when they heard recorded playback of cahow courtship calls, and one couple was captured and physically moved there! Three pairs nested in the new burrows.

In the early spring of 2008, I was able to speak to another dedicated advocate for the cahow, Jeremy Madeiros. Jeremy first became involved in 1984 in his late twenties when he was accepted as a training apprentice under David Wingate in what was then the Department of Agriculture and Fisheries. As a boy, Jeremy had preferred poking about for insects and plants over kicking balls around with his friends. The experience he gained working with David—not only on the recovery of the cahow, but also in the efforts to restore Nonsuch Island as a new nesting ground for the species—was just what he needed. He went to college and got the qualification that eventually landed him a job as a parks superintendent. He was able to maintain his connection with David as he followed in his footsteps.

Learning to Live with Danger

Above all, Jeremy needed to learn to cope in often dangerous conditions—"to work without killing or injuring myself," is how he put it during a long telephone conversation we had. Knowing that David took huge risks, I asked Jeremy what it had been like to work with him. He laughed and told me about something that happened in the early 1990s when the two of them were monitoring the progress of the cahow chicks. This is done at night, when the chicks come out of their nest burrows to explore and stretch their wings. David decided to start the monitoring on an islet where they knew there were two nests. With only the light from their flashlight (the chicks do not come out in moonlight, which would make things so much more convenient for humans), they had to maneuver the little boat close to the rocky shore in a high swell.

"We had to jump onto a rock and quickly scramble up before the next wave covered it," said Jeremy. Then they had to get to the far side of the islet, which meant climbing a steep cliff as there was no access by boat. They got there safely and, as always, had a wonderful time watching the chicks. It was on the way back that disaster so nearly struck.

"David was having trouble with his back," Jeremy told me, and had taken a foam rubber cushion to sit on the sharp rocks. At one place they had to jump down ten feet onto a rock below—a rock that was flanked, on either side, by a twenty- to thirty-foot drop onto jagged rocks and crashing waves.

"He asked me to go first," said Jeremy, "and then he threw down the cushion and asked me to put it on the rock. He thought it would lessen the jarring to his spine." Imagine Jeremy's utter horror when David landed safely—only to bounce right over the edge and vanish from sight. "I hardly dared shine my torch down," Jeremy said, "I was so sure I would see a mangled body way down below." How could anyone survive such a fall? And if he had survived, how could he, Jeremy, possibly get there with the boat to rescue him?

"I nervously shone the torch down," Jeremy said, "and there were two eyes looking up at me." Somehow David had managed to grab onto a jagged rocky outcrop. He was battered and bloody but very much alive, and with Jeremy's help he managed to scramble back up. And insisted they visit the other chicks on their list!

A New Home for the Cahow

After Hurricane Fabian destroyed so many cahow nesting sites, it became clear that the birds' long-term survival would depend on the restoration of some of their original nesting habitats. And this is where the future of the cahow becomes linked with David's extraordinary restoration work on Nonsuch Island (described in the sidebar). When the time came to start a new colony of cahow on the restored island, a blueprint for the translocation of petrel chicks already existed: Nicholas Carlile and David Priddel had successfully established a colony of endangered Gould's petrels on a new island—the whole fascinating story is told on our Web site (janegoodallhopeforanimals.com).

"We could not have risked trying relocation with the cahow if we had not known of the success of Nicholas's work with the Gould's petrel," David told me. "The cahow were still in such a precarious state."

In 2003, Nicholas joined the cahow restoration project. He assisted in designing a recovery plan that had the ambitious goal of moving a hundred young birds to Nonsuch Island over a five-year period. The first of these translocations was made that year—ten chicks, three weeks before fledging, were taken from their nests on the islets to artificial burrows constructed for them on a by-then rat-free Nonsuch. They were fed each night, and their growth and behavior recorded.

Nicholas had found that it is very important not to move chicks too late. It is when they first leave their nests to look around (about eleven days prior to fledging) that the location of the nest is imprinted in the brain, so that it will be that place—rather than the spot where they hatched—to which they will subsequently return, three to five years later, to nest themselves.

When those first chicks were moved, Jeremy worried a bit. They were going from bare rocks to wooded slopes—would they be able to cope?

"Nicholas was there when we moved the first youngsters," Jeremy told me. "We were amazed as we watched a chick emerge from its nest burrow, stretch its wings, and move around exploring. Suddenly it came to a tree. It stopped, looked up—and immediately scurried right up the trunk like a squirrel, using its sharp little beak and claws, and sort of hugging the trunk with its wings. Right on up to the top!" Of course, when they thought about it, this made sense. Tree climbing is probably deeply encoded in the birds' ancestral memory, for in the old days, emerging from their burrows in the forests, they would have climbed in order to take off to sea from the treetops. Since then the poor things have been reduced to climbing up bare rocks.

"After that," said Jeremy, "I realized why the chicks on the islets so often climbed up David and me and fledged from the tops of our heads. We were the closest thing to a tree in their abnormal world of rock!" Jeremy paused and laughed. "They often left their mark on our heads before they left," he said. "But that was okay—it's supposed to be lucky!"

All of the first ten translocated chicks fledged successfully and flew off to spend the next few years at sea. The following year, twenty-one were moved, and again they all fledged successfully. Just before the 2008 breeding season, eighty-one out of the planned hundred have been successfully moved, and seventy-nine of them have fledged and departed safely.

Stop-Press News

Recently I received news from Jeremy. "I said I would let you know if anything exciting happened," he wrote. "I am happy to report (with a big smile on my face!) that just such a thing has now happened!"

But first he reported on the situation of the original four tiny breeding islets where the population is continuing to grow. Originally there were just eighteen breeding pairs—but the number has now risen to eighty-six. "It seems that, perhaps because the colony is getting bigger—which they love—there is more pair formation going on," said Jeremy. "It is as though they have changed to a higher gear. And once the critical mass has been reached, there will be more and more pairs each year. Then they will be on their own."

When he wrote, Jeremy was busy checking the weight, wing growth, and plumage development of the forty chicks hatched on the islets in 2008, twenty-one of which will be moved to Nonsuch. And if all twenty-one fledge successfully, this will mean that their goal will have been reached: One hundred cahow chicks will have been moved to and fledged from Nonsuch during the first five years of the project.

Next, Jeremy shared his really exciting news. In mid-February 2008, he was on Nonsuch carrying out some repairs to the solar-powered sound system that has been installed at the new nest site. It plays back courtship calls to encourage any cahow within hearing distance to investigate. Jeremy decided to stay overnight on the island to see how it was working.

"About forty-five minutes after nightfall," Jeremy told me, "the first cahow swept in from the open ocean and started circling above the translocation site; more came in and began to carry out acrobatic high-speed courtship flights until within another hour I could see a maximum of six to eight birds at once. Sometimes they circled high above; sometimes [they] made low, acrobatic high-speed courtship flights just above the artificial nest burrows, often making their eerie moaning calls."

Eventually, some of the birds began to land among the burrows, "culminating with one bird landing right beside me! I was able to just reach right over and gently pick it up without any fuss." Jeremy confirmed from its band number that it was indeed a bird that had been moved to

Nonsuch as a chick in 2005. "My heart just leaped when I realized that this bird had not just survived the last three years at sea after being partly raised by us, but had in fact returned to its point of departure as hoped!"

More cahows were recaptured at the site over the next month or so, and all were the birds that had been translocated. In mid-March, one of them was found for the first time staying for a whole day in one of the Nonsuch burrows, excavating a large pile of soil outside the nest entrance, digging a nest scrape in the nest chamber, and pulling in nest material. "A sure sign that this bird has now 'claimed' this burrow," said Jeremy. I could hear his excitement as he told me that he had checked its band and found that this was the exact same burrow to which it had been moved in 2005! "And," he said, "I had watched it fledge to sea during a night watch in June 2005. How amazing to think it has carried out a perfect 'return to the point of departure' after living God knows where out on the ocean!"

In all, four cahows translocated to Nonsuch in 2005 were captured near the nest burrows. Between six and eight were observed some nights flying over the site; at least six nest burrows received prospecting visits, some more than half a dozen times. And cahows have stayed over for the day in three of these burrows on several occasions. Jeremy thinks those birds were probably males, which seem to return a year or two earlier than the females, and he hopes that next season they will return and start attracting females to their burrows. "And by then, the first of the 2006 translocated cohort should also be joining them. I can hardly wait!"

When the cahow that fledged on Nonsuch return to breed there themselves, it will be a major milestone in the restoration of this resilient seafaring bird and a tribute to the determination of Jeremy Madeiros, Nicholas Carlile, and, above all, David Wingate, who fell in love with the cahow as a schoolboy fifty-nine years ago.

NONSUCH ISLAND

Situated off the coast of Bermuda, Nonsuch Island has a strange and altogether fascinating history. In 1860, the British colonial government wanted to establish a yellow fever quarantine station. So it bought the tiny Nonsuch Island (less than fifteen acres and sixty feet at its highest) from a private owner who had been using it for grazing cattle.

The quarantine station and hospital that were built served their purpose for fifty years before it was decided, for logistical reasons, to move the operation to Coney Island. Soon after this, in 1928, the island was loaned to the New York Zoological Society for use as a marine research station. Then, in 1934, Nonsuch became, of all things, the site of a training school for delinquent boys. But in 1948, because the island was so very isolated, and because of its rocky shoreline that made access really difficult, the school was moved elsewhere.

For the next three years, the little island was left to itself. By this time it had become a rather sad and barren place, for an epidemic of a juniper scale insect had destroyed almost 95 percent of the forest that previously covered the Bermuda islands—and Nonsuch was virtually denuded. Then, in 1951, something happened that would utterly change the future of Nonsuch. A small colony of cahow was rediscovered breeding on a couple of offshore rocky islets. And it became apparent that the birds would soon be truly extinct if they did not have a more appropriate habitat for breeding. Nonsuch Island was, it was thought, ideal—for cahow had bred there before. But first its damaged environment would have to be restored

In 1962 David Wingate, who years before as a sixteen-year-old schoolboy had been with the group that made the cahow discovery, moved onto Nonsuch Island as a warden. This was the start of the extraordinary restoration project, which was the focus of David's career for the next forty years.

More than eight thousand seedlings of native tree species, some of them endemic to Bermuda, were planted, along with two fast growing non-native species—the Australian casuarina and European tamarisk. These were used as a stopgap measure to replace the windbreak lost after the endemic cedars had been killed during the juniper scale insect epidemic. Over the next twenty years, the upland forest became well established, and when Hurricane Emily hit the island in 1987 it caused little damage to the endemic and native trees. As the forest thrived, the non-native trees were gradually ring-barked—a thin strip of bark was removed around the base of each tree so that it died slowly, causing minimal disruption.

Meanwhile another major project began in the mid-1970s, when two small artificial ponds were constructed to re-create saltwater and freshwater marsh habitats. Nicholas Carlile, who has several times visited Nonsuch, told me that it was truly amazing—on one tiny island of just fifteen acres "they have re-created several complete ecosystems," including the rocky coast, coastal hillside, marshes, upland forest, and beach dunes.

Many of the plants now flourishing on Nonsuch are endangered on Bermuda's main islands, where approximately 95 percent of the total biomass is exotic. The Nonsuch project was one of the very first to involve restoration of an island on which virtually all of the flora and fauna had been totally eliminated by human degradation or invasive pests. The extraordinary success resulted from taking a holistic approach: eliminating the pests and restoring the entire terrestrial ecosystem as close as possible to its original state. It was the success on Nonsuch Island that led to other restoration projects on other islands as far away as New Zealand.

Once the habitat had been restored, it was possible to use Nonsuch as a reintroduction point for a variety of species including a night-heron, the West Indian top shell, and green turtles, all of which have been locally extinct in Bermuda for a hundred or more years. Out of one man's dreams and determination sprang the "living museum" concept that has inspired the transformation of Nonsuch Island. It now presents an almost true replica of the prehistoric native environment of Bermuda and its islands before humans destroyed so much. From the outset, it was David's ultimate goal "to create on Nonsuch an optimal habitat for Bermuda's banner species and national bird—the burrow-nesting cahow." As we have seen, that ultimate goal was reached.

The name of Carl Jones is synonymous with restoration of endangered species on Mauritius. Shown here with the dazzling, emerald-green echo parakeet, the last of perhaps as many as seven parakeet species once found on the islands of the western Indian Ocean. (*Gregory Guida*)

The Birds of Mauritius

Mauritius Kestrel *(Falco punctatus)*
Pink Pigeon *(Columba*
[formerly *Nesoenas*] *mayeri)*
Echo Parakeet *(Psittacula eques echo)*

When I think of these birds, I think of Carl Jones. If he had not gone to Mauritius (an island nation off the coast of Africa), it is more than possible that all three species would be extinct, for he has led the fight to save them—even when, at times, it must have seemed a daunting, if not impossible, task.

It took me some time to track Carl down at his home in Wales, where he spends time when he is neither in the field nor at the offices of the Durrell Wildlife Conservation Trust in Jersey. We had a long conversation by telephone, and although I would have preferred to meet him in person, Carl's warmth and his love for his work are so genuine, his enthusiasm so infectious, that I feel I have known him for a long time. I learned that he is very interested in bird psychology and that he lives on a small holding with his family that includes some parrots, an eagle—and a tame condor that is imprinted on humans and treats Carl as his partner! Carl told me that he shares my belief that not only is it okay for a scientist to feel empathy with the animals he studies, it's in fact necessary for real understanding.

The three stories I want to share together represent a heroic struggle, ultimately successful, to save three very different species from extinction—a falcon, a pigeon, and a parakeet. By the late 1970s when Carl stepped in, all three of these species had been critically endangered for many years and were on the very brink of extinction: There were only four Mauritius falcons in the world, only ten or eleven pink pigeons, and around twelve echo parakeets.

The Mauritius Kestrel *(Falco punctatus)*

Carl's fondest memories are of the many seasons that he worked with Mauritius kestrels in their last home, the Black River Gorges. At that time, he told me, much of his life revolved around this small, charismatic falcon. It is just under a foot in length, with the male weighing only about 4.7 ounces—smaller than the 6.3-ounce female. They have pure white underparts with round or heart-shaped spots. "For me," Carl said, "they were the most beautiful of birds, and I used to get very excited when I just got a glimpse of one. They have distinctive rounded wings and are very maneuverable. They weave in and out of the forest canopy chasing and feeding on the bright red and green day geckos that are their main prey.

"They used to ride the updrafts from the sides of the cliffs, rising hundreds of feet, and then just close their wings to plummet earthward, hurtling vertically downward at great speed," he continued. "Sometimes they would pull out of their stoop and just land gently on a tree or on the cliff; more usually they used the momentum to shoot upward again."

As the breeding season approached, they became more and more aerial, Carl told me. "They would chase each other around and fly in the most beautiful 'sky dances,' rising and falling in gentle undulations or in jagged zigzags. Often they would just rise in the sky on a thermal, flying around together and calling until sometimes this courtship display culminated in mating in their nest cavity." Although Carl was talking of his experiences of some thirty years ago, he told me, "I cannot think about these early observations of the kestrels without a flush of excitement and a quickening of the pulse."

Teetering on the Brink

The Mauritius kestrel had been pushed to the verge of extinction as a result of severe deforestation during the eighteenth century—accelerated by the devastating effects of cyclones, predation from invasive species (especially crab-eating macaques, mongooses, cats, and rats), and the 1950s and 1960s use of pesticides, especially DDT, for malaria control and food crop protection.

In 1973, the Mauritius government had agreed to the capture of one of the last pairs of these falcons for an attempt at captive breeding—

which failed. One chick was born but it died when the incubator broke down, and subsequently the female died as well. By the following year, there were only four Mauritius kestrels remaining in the wild, and it was considered the rarest bird in the world.

It was in 1979 that Carl started his work on Mauritius, under the auspices of the Durrell Wildlife Conservation Trust. He became the sixth biologist in as many years to work with the kestrels. Although he was only twenty-four years old, he had spent many years keeping and rehabilitating injured birds. Fresh from college with a degree in biology and knowledge of recent advances in breeding falcons in captivity, he had, he told me, "the enthusiasm and arrogance of youth." He had seen breeding success with injured common kestrels in his parents' garden and was sure that he could save this the rarest of all birds where others had failed.

The Hazards of Egg Snatching

Carl knew that common kestrels, like so many birds, will lay again if their first clutch is removed, and he determined to try the technique on the wild Mauritius kestrels. He had to climb steep cliffs to reach the "nests"—shallow depressions, or scrapes, in the substrate—of each of the two breeding pairs to take their eggs.

"The first nest was on a relatively small cliff, and I could get to it by using an extension ladder," he said. "I found that the kestrel had laid its eggs at the back of a small cave about two meters deep. I crawled in and took the three eggs and placed them carefully in a widemouthed insulated flask that had been preheated to the correct incubation temperature." From there they went to an incubator at the government's captive breeding center, about five miles away.

The second nest was on a high cliff, and Carl had to be lowered down to it on a rope. "The eggs were deep in a narrow cavity that opened up into a nest chamber about four feet into the rock, and the only way I could reach them was by attaching a spoon to a long stick. The eggs had been laid in the remains of a dead tropicbird and were in a bed of soft white feathers." Those eggs soon joined the others in the breeding center.

Because the species was so close to extinction, this was a tense time, and Carl camped on the incubator room floor to be close by in case anything went wrong. Four of the eggs hatched, and he hand-reared the chicks "on minced mouse and minced quail." All four fledged, and since the double-clutching technique had worked so well, it was repeated in subsequent years. Thus a captive population was built up, and these birds subsequently bred successfully. Gradually the total number increased.

In 1984, Carl took a chick from the captive breeding center and put it in the nest of one of the wild kestrels, Suzie. She reared it successfully, and it became the first captive-born individual to return to freedom. Subsequently captive-bred and -raised birds were released into areas where there was suitable habitat but no kestrels.

In 1985, Carl was able to announce the fiftieth successful hatching at the breeding center from captive-laid and wild-harvested eggs. And by 1991, as a result of double-clutching in the wild and captive populations, artificial insemination, and successful raising of incubator-hatched chicks, two hundred Mauritius kestrels had been successfully bred. By the end of the 1993–1994 breeding season, 333 birds had been released to the wild.

Meanwhile Carl and the DWCT, working with the Mauritius government, were continuing their work with the wild population. Supplementary food was provided, and the birds were offered—and used—nest boxes. Strict predator control served to reduce numbers of introduced predators, and work on habitat restoration was begun. This meant that captive-bred and -reared birds released into the wild had a good chance of survival. Indeed, in the early 1990s the kestrel population was judged to be self-sustaining, and, said Carl, "the captive breeding program was closed down, the job was complete, and the kestrel was saved." Indeed, recent studies have shown that there are probably more than a hundred breeding pairs and a total of about five to six hundred birds. Kestrel lovers—raise your glasses to the success of this effort!

The Pink Pigeon *(Columba* [formerly *Nesoenas] mayeri)*

Most people think of pigeons as pests. We all know the overfed birds that strut unconcerned along the pavements of busy cities, congregate around people eating in the park, and deface the walls of buildings on which they roost. Forget all that. The pink pigeon is a beautiful, medium-size pigeon with a delicate pink breast, pale head, and foxy red tail.

"This stunning bird," said Carl, "had been rare for probably two centuries or more and for a while was thought extinct." Then in the 1970s, a tiny population of about twenty-five to thirty birds was found surviving in a small grove of trees high on a mountainside that had one of the highest rates of rainfall in Mauritius—about fifteen feet per year. They lived there, Carl told me, not because they liked it but because the number of predators was low in this wet and often cold habitat. But even there their numbers were declining due to habitat destruction and degradation, and because of introduced monkeys and rats that raided the nests and ate the eggs and young. Feral cats killed adult birds as well.

By 1990, there were only ten or eleven known individual pink pigeons left in the wild, and it appeared that the tiny population was in terminal decline. Fortunately in the mid-1970s, a team from the Durrell Wildlife Conservation Trust had captured a group of pigeons for a captive breeding program run by Carl. He had studied this group for his PhD degree.

"They were a real challenge to breed," he told me. "They were very fussy about their mates, and to find compatible pairs was a real headache." With small populations, of course, it is important to manage the genetic diversity and to prevent the mating of closely related individuals. But, said Carl, "It was common for the birds to reject the partners that you felt were most appropriate and then try and pair up with their first cousin or even a sibling! Sometimes I felt like a pink pigeon marriage guidance counselor . . . a compatible breeding pair might breed and then one day there would be a huge bust-up and one would be beating up the other and they would have to be separated."

Despite the problems, the pigeons started breeding. But then they proved to be such poor parents that the eggs and young had to be reared under domestic doves. In time, however, by allowing them to practice rearing young doves, Carl was able to improve their parental skills. And so, finally, with the pink pigeons breeding and raising their young at Black River, Carl and his team developed a program for releasing them back into their native forest.

Under Carl's supervision, a young Englishwoman, Kirsty Swinnerton, pitched a tent in the forest and monitored their progress for five years. It soon became obvious that they faced a variety of problems. First, especially at certain times of the year, there was very little appropriate food in the forest, much being eaten by introduced monkeys, rats, and birds. This meant that supplementary food needed to be provided. Second, when the reintroduced pigeons started to breed, several of them were killed by feral cats, necessitating increased predator control. But when these problems had been addressed, the original released population gradually began to increase so that eventually it was possible to establish several other populations. And in 2008, Carl told me, there were nearly four hundred free-living pink pigeons divided among six different populations. "This species is now secure," he said.

The Echo Parakeet *(Psittacula eques echo)*

Having attained considerable success with the Mauritius kestrel and pink pigeon, Carl turned his attention to what was then the world's rarest parrot—the beautiful emerald-green echo parakeet. It is the last of the three or four species of parrots that once lived on Mauritius, and the last of perhaps as many as seven parakeet species once found on the islands of the western Indian Ocean.

In the 1700s and early 1800s, the echo parakeet was very common in Mauritius and Reunion Island in upper- and mid-altitude forests and in the scrublands—the so-called dwarf forest—feeding on fruit and flowers in the upper branches and nesting in holes in the trees. The Reunion population disappeared first, and between the 1870s and the 1900s the population in Mauritius gradually fell. This was due primarily to habitat loss and competition from introduced species. Fortunately in 1974, and as a result of growing awareness, the remaining forest was given almost total protection, and a significant nature reserve was created by linking smaller protected forests. But for a while, it seemed that this move had come too late—the tiny population of echo parakeets was having limited nesting success.

In 1979, when Carl was spending a lot of time in and around the Black River Gorges with his kestrels, he occasionally saw small flocks of the parakeets on the ridges surrounding the gorges. They were, he said, tame and confiding, and because they sometimes fed only a few feet away from him, he got to know them individually. But they were disappearing fast: By the 1980s, there were only eight to twelve known individuals left, of which only three were females—although Carl says it is possible that several birds had been overlooked.

Since these parakeets were island residents, facing similar problems to the birds of New Zealand, Don Merton was invited to help the effort to save them from extinction. Drawing on his considerable experience and working closely with Carl, he devised and helped implement the recovery strategy. First, they initiated a study to get to the bottom of the parakeets' nesting problems. They found that when the parakeets did breed, the chicks were attacked by nest flies that would in some years kill most if not all of them. This meant that nests had to be treated with insecticides. Another problem was tropicbirds taking over nest sites, so tropicbird-proof entrances had to be installed on suitable nest cavities. Rats also posed a great threat, sometimes eating both eggs and young. After two precious nests were lost to rats, the team stapled rings of smooth PVC plastic around the trunks of each nest tree and placed a bucket of poison nearby. One nest was attacked by a monkey, who grabbed a chick and wounded the mother. The team isolated nest trees by judicious pruning of the canopy so that monkeys could no longer jump in from neighboring trees. Then there were the seasonal food shortages—and so feeding hoppers were introduced (though it was many years before the birds learned to use them). Finally, nest cavities were made more secure and weatherproof.

The biologists found that though females typically laid three or four eggs, usually only one chick fledged. In other words, chicks were dying in almost all nests. Carl and his team decided that if there were more than two chicks in a nest, they would take the "surplus," leaving the parents with a brood they could raise comfortably. If a pair failed to hatch any eggs, a "surplus" chick was given to them from another nest.

"In such intelligent birds as the echo parakeets," Carl told me, "it is important for their psychological well-being that they are allowed to rear young. It is also important for the young to be reared in family groups." This program of manipulation of nests also resulted in many surplus young being taken to the breeding center, where they were raised successfully.

The first three captive-bred birds were returned to the wild in 1997; others soon followed. But there were problems with these hand-reared birds. "Some were just too tame," Carl told me. "When they saw you in the forest, they would fly down and land on your shoulder." And they were very naive. Sometimes they landed near a cat or mongoose—and did not live to tell the tale. Carl spent a lot of time with these young birds, pondering their problem. He had been releasing them when they were seventeen weeks old, so he decided to try releasing the next youngsters at about nine to ten weeks—the time when they would normally fledge. The results were dramatic. "These younger birds integrated with the wild birds and learned their survival and social skills."

Gabriella was one of the first three birds to be released. She mated with a wild male, Zip, and was the first captive-bred female to fledge a chick—Pippin. Gabriella had learned to use a feeding hopper in captivity and Zip, learning from her, became the first wild bird to use one.

In subsequent years, the number of birds taking supplemental food from the hoppers and using nest boxes provided by the team gradually increased, as did the number of breeding pairs. By 2006, it was decided to stop the intensive management of the wild birds, only continuing with the supplementary feeding and provision of nest boxes. In March 2008, I learned that there are about 360 free-living echo parakeets—and the population is still growing.

A Haven for the Future

And so, the echo parakeet represents another species saved—although it will be necessary, said Carl, to continue with supplemental feeding and predator control. Skeptics maintain that a species cannot be deemed secure until it can live on its own, independent of human help. "But," said Carl firmly, "in an increasingly modified world, we are going to have to look after and manage the wildlife if we want to keep it." Alas, he is right. In a world so damaged by our human footprint, it is likely that we shall have to remain eternally vigilant to protect threatened and endangered species: They need all the help we can give them. It is the least we can do.

One of the most important projects on Mauritius, along with ongoing predator control, is the restoration of areas of native forest—a program in which the government's National Parks and Conservation Service now plays a large part. As a result of the successes with the Mauritius kestrel, pink pigeon, and echo parakeet, the prime minister of Mauritius declared the Black River Gorges and surrounding areas Mauritius's first national park—a haven "for the birds that have been saved to live."

Hawaiian Goose or Nene (*Branta sandvicensis*). Coming home: A mated pair (male left, female right) of Nene, shown here overlooking the eruption at Hawaii Volcanoes National Park in 2006, have begun to return to their original lowland habitat. (*Niki Endler*)

Cotton-Top Tamarin (*Saguinus Oedipus*). There are probably less than ten thousand tamarins remaining in their natural habitat of northwest Colombia. In the last century these little monkeys were over-collected for biomedical research into colon cancer. Today their biggest threat is habitat loss, which is why the community efforts to protect them and their forest habitat are so vital. (*Proyecto Titi, Inc.*)

Panamanian Golden Frog (*Atelopus zeteki*). One of the many amphibians endangered worldwide because of the lethal chytrid fungus. In an effort to protect this beautiful amphibian from extinction, conservationists set up a "Frog Hilton," a quarantine and breeding program literally inside a hotel in Panama. *(William Konstant)*

Abbott's Booby (*Papasula abbotti*). These big birds have big personalities, according to a couple who feed and nurture wounded and orphaned boobies. The birds' survival depends on the ongoing restoration of Australia's Christmas Island National Forest. (*Dr. Janos Hennicke*)

Bermuda Petrel or Cahow (*Pterodroma cahow*). These mysterious, exotic birds spend up to three years at sea before coming home to nest—and no one knows where they go. Shown here, a cahow soaring over the ocean at dusk near Nonsuch Island, Bermuda, where heroic (even life-threatening) restoration efforts are saving these glorious birds from extinction. (*Andrew Dobson*)

Mauritius Kestrel (*Falco punctatus*). This small, charismatic falcon was considered the rarest bird in the world in the 1970s after being devastated by deforestation, cyclones, invasive species, and DDT. Biologist Carl Jones, who had seen breeding success with injured common kestrels in his parents' garden, figured out a fascinating way to save this species where others had failed. They are now safe and self-sustaining, with 500-600 birds living free on Mauritius Island. (*Gregory Guida*)

Pink Pigeon (*Nesoenas mayeri*). By 1990 there were less than a dozen known individuals left in the wild, and it appeared that the pink pigeons of Mauritius were doomed. Amazingly, after surmounting many challenges with captive breeding (which even required some pigeon marriage counseling), the species is now secure, with nearly four hundred thriving in the wild. (*Gregory Guida*)

Echo Parakeet (*Psittacula eques*). Once considered the world's rarest parrot, the glorious echo parakeet has been saved and restored on the island of Mauritius thanks to biologist Carl Jones and the Durrell Wildlife Conservation Trust. (*Gregory Guida*)

Short-Tailed Albatross or Steller's Albatross (*Phoebastria albatrus*). Adult short-tailed albatrosses in a courtship display—they lightly touch bills in a sort of "fencing"—in their nesting grounds on Torishima, Japan. This amazing seabird was once down to less than a hundred survivors. (*Hiroshi Hasegawa*)

Blue and Gold Macaw (*Ara ararauna*). Poaching for the illegal pet trade extirpated this beautiful parrot from its native habitat on the island of Trinidad. When the release of captive-bred birds didn't work, the people of Trinidad figured out an innovative way to rescue and restore their beloved blue and gold macaw to its home in the Nariva Swamp. (*Bernadette Plair*)

Black-Crowned Dwarf Marmoset (*Callithrix humilis*). This tiny primate is one of six new marmoset species recently discovered in the depths of Brazil's Amazon forests. I met one of these adorable primates who'd been rescued from a local village. As she sat on my shoulder, I wondered how many of her kind are still out there, living undiscovered and needing our protection. *(Russell A. Mittermeier)*

Yariguies brush finch (*Atlapetes latinuchus yariguierum*). It's comforting to know that there are still hidden, remote places where animal species are living unseen by humans and still "undiscovered" by science. This brush finch, discovered in Colombia in 2007, was the first bird species from the New World that was deliberately not killed to provide type specimens. *(Blanca Huertas)*

Tahina Palm (*Tahina spectabilis*). Even as we face the "sixth great extinction," we are simultaneously making new discoveries on Earth. Imagine the amazement of Xavier Metz, the manager of a cashew plantation, when he discovered this giant palm—not only a new species, but a new genus, from an evolutionary line that was not known to exist in Madagascar. (*John Dransfield*)

"Ant from Mars" (*Martialis heureka*). This blind, subterranean, predatory ant was discovered in an Amazon forest in 2008. It is likely a direct descendant of one of the very first ants to evolve on Earth over 120 million years ago. (*© 2008 National Academy of Sciences, U.S.A. originally appeared in PNAS vol. 105, no. 39: 14913 – 14917*)

Kipunji (*Rungwecebus kipunji*). This monkey was discovered in 2003 in the southern highlands of Tanzania—one of the exciting new species that are still being found in remote areas. The kipunji is not merely a new species, but a completely new genus—a kind of sister to baboons. (*© Tim Davenport/WCS*)

Coelacanth (*Latimeria chalumnae*). This enormous, highly elusive fish was considered extinct for some 65 million years until its rediscovery in 1938. This image was extracted from an extremely rare videotape by the Coelacanth Diving Team in 2000, which found the creature at 108 meters in Jesser Canyon, Sodwana Bay, South Africa. *(Pieter Venter and the Coelacanth Diving Team)*

Crimson Spider Orchid (*Caladenia concolor*). There are only eighty known individuals left in the native habitat of Australia's Box-Gum Woodlands. Thanks to the vigilance and protection of local citizens, including the indigenous Wiradjuri community, the orchids and their habitat are now protected and gradually increasing in numbers. *(Robert G. Fleming, Wagga Wagga)*

Mauna Loa Silversword (*Argyroxiphium kauense*). These striking silverswords of Hawaii were almost lost, but for the dedication of field botanists such as Robert Robichaux. He told me a harrowing story about dangling from a rope above a 50-foot drop so he could hand-pollinate a few remnant silverswords that grow on a cliff. (How much easier for a bee with wings!) (*Silversword Foundation*)

Sudbury, Ontario, before and after: This area was once ravaged by logging and copper mining. These photos, taken from the same place, clearly show that determination and persistence can heal habitats. (Note the white tower on the right in both photos.) *(City of Greater Sudbury)*

Paradise restored: After the invasive animal species were removed from Mexico's Guadalupe Island, the indigenous species like this *Guadalupe sencios* could finally thrive once again. *(Claudio Contreras Koob)*

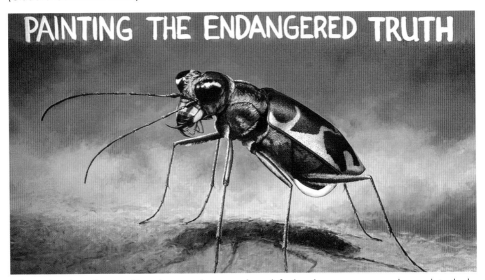

When this beetle was listed as endangered and federal money was released to help safeguard its unique salt creek habitat in Nebraska, there was an outcry of local protest. Yet, in the web of life, all species and habitats are crucial. *(Jessa Huebing-Reitinger)*

Short-Tailed Albatross or Steller's Albatross
(*Phoebastria albatrus*)

The story of the short-tailed albatross is inexorably linked with one man, Hiroshi Hasegawa, and his lifelong dedication to a single cause—saving an extraordinarily beautiful and extremely endangered bird from extinction. This bird made its last stand in a remote and almost inaccessible corner of the world—Torishima, an active volcano island that rises in sheer and mostly unscalable cliffs out of the sea, some eleven hundred miles southeast of Tokyo.

I spoke with Hiroshi during my annual visit to Japan in November 2007. I was very excited to meet this extraordinary man. His eyes are bright with love for his work, and for the birds to which he has dedicated his life, and he seems filled with suppressed energy. I longed to go with him to watch the short-tailed albatross—but I must make do with the information he has so generously shared with me.

Growing up in the hilly mountainous area near Fuji, he developed a passion for birding that eventually led to his love for the short-tailed albatross, the largest seabird in the North Pacific. Their long narrow wings, with a span of more than seven feet, enable them to glide effortlessly, low over the ocean, going ashore only during the breeding season between November and March. They are very beautiful; the adult has a white back, golden-yellow plumage on the head, and black-and-white wings. Most distinctive is the bill, which is long and bubblegum pink, tipped with blue.

At one time the short-tailed albatross was common, ranging for miles from Japan to the West Coast of the United States and the Bering Sea and nesting on grassy slopes set among the rocky cliffs of small islands,

A chick begs its parent for food on Torishima Island. When Hiroshi Hasegawa first set foot on this island in 1977 he found only fifteen struggling chicks among the seventy-one surviving albatrosses—and he knew then that these beautiful birds were on the verge of extinction. *(Hiroshi Hasegawa)*

mostly off Japan. It was their glorious plumage that almost led to their extinction: Between 1897 and 1932, it is estimated that feather hunters clubbed to death at least five million of them on their main breeding grounds on the rugged cliffs of Torishima. By 1900, there were some three hundred feather hunters camped there during the breeding season, and the numbers of short-tailed albatrosses continued to decline. When the hunters heard that the Japanese government, in response to lobbying from ornithologists and conservationists, had agreed to make the island off limits, they organized a final massacre. At the end of the slaughter, no more than fifty individuals remained. And then, in 1939, another volcanic eruption wiped out most of the last nesting sites.

At least the few survivors now had legal protection: The Japanese government had listed the short-tailed albatross as a Special National Monument, as well as protecting Torishima Island as a National Monument. But there were very few left to protect—in 1956, an expedition counted only twelve nests. Seventeen years later, British ornithologist Dr. Lance Tickell went to Torishima Island to check on this tiny colony and to band the chicks. On his way back, he stopped to give some lectures in Japan's Kyoto University. That visit made a deep impression on Hiroshi Hasegawa, then a graduate student majoring in animal ecology. Indeed, it determined his future. If a British ornithologist could get to the remote Torishima Island, in Japanese waters, then surely he, Hiroshi, could somehow get there himself.

He could hardly have set himself a harder task. For one thing, he had no funding. And when he eventually got a place on a fisheries research vessel going to Torishima, the weather was too bad for them to land and he only glimpsed the nesting albatrosses from the ship.

Finally, in 1977, Hiroshi set foot for the first time on Torishima Island. He counted only seventy-one adult and immature birds. Since the short-tailed albatross probably lives to be fifty or sixty years old, some of the adult birds were almost certainly survivors of the 1932 massacre. There were only nineteen chicks among the seventy-one birds—four of them already dead, while the other fifteen died before fledging. Hiroshi knew, then, that these beautiful birds were very, very close to extinction. "I understood," he told me, "that it was my responsibility, as a Japanese, to bring the species back from the brink."

For a while, Hiroshi was supported by a fisheries experimental station, but their boat had an annual schedule that was not geared to the breeding season of the albatrosses. He succeeded in getting funding for a few years from the Ministry of Education, Science and Culture, but the government would not commit to the long-term project that Hiroshi knew was necessary. And so, he told me, he gave up seeking funding from official sources and instead began writing a series of popular articles and children's books. This brought in sufficient funds to charter boats when he needed them for his albatross work. It was then that he learned "never to copy others' ideas." Instead he developed his own vision of a conservation plan.

A Rare Bird and a Rare Man

The journey to the breeding grounds is tough. First comes a long boat ride over the open sea—and there can be horrific storms. Even ashore, all the equipment must be hauled up sheer black volcanic lava, to a height the equivalent of fourteen stories, and then down a four-hundred-foot cliff before arriving at the breeding site. Hiroshi has made this journey two or three times a year for twenty-seven years. All the more remarkable considering that, as he confided to me, he always gets seasick! During the breeding season from early November to late December, Hiroshi counts birds and nests on the island, and observes their behavior. In late March, he returns to put identifying bands on the chicks' legs. And in June, he sometimes goes back to work on improving the nesting sites, planting grass to stabilize the soil and provide some cover. Gradually, the survival rate of the chicks increased. But in 1987, probably as a result of a fierce typhoon and very heavy rain, there was a massive landslide on Torishima Island, followed by a series of bad mudslides that destroyed some nesting sites. This probably caused increased competition for space with black-footed albatrosses.

Hiroshi realized then that it was desperately important to establish a new nesting colony in another part of the island. He carved life-like decoys (to date he has produced about one hundred), which he placed at the site he had selected. Then, when the adult birds began returning for the breeding season, he played back courtship calls of short-tailed albatross (a method pioneered by Dr. Steve Kress when working with Atlantic puffins). For the first two years, there was no response. Then, for the 1995–1996 breeding season, one pair nested there and successfully reared a chick. No other individuals arrived the next year, nor the one after that, but Hiroshi did not give up. He continued to put out decoys and play calls, year after year, until finally, ten years after the first pair had raised their chick, three more pairs arrived. By the 2006–2007 breeding season, the new colony numbered twenty-four nesting pairs; sixteen chicks were fledged.

Meanwhile the breeding success at the original site gradually improved. In the 1997–1998 season, 129 chicks fledged (67 percent of all those hatched); the following year, 142. And so it went, year after year, until during the 2006–2007 breeding season no less than 231 chicks fledged, and the population of the colony was almost 2,000. One of these is a bird banded by Tickell that Hiroshi has been observing since the start of his study; it successfully reared a chick at the age of thirty-three years.

Threats at Sea

Of course, short-tailed albatrosses—like all the albatross species—face major threats during their months at sea. Many are hooked and drowned on commercial long lines; others get tangled in abandoned fishing gear or swallow plastic debris floating in the ocean. From time to time, they are coated with oil from spills. Hiroshi and other ornithologists tried to raise public awareness. Between 1988 and 1993, a series of TV programs about the plight of the short-tailed albatross was broadcast throughout Japan. In 1993, the short-tailed albatross was listed as endangered in the Japanese Endangered Species Act. And finally, nearly twenty years after beginning his battle to save these birds, Hiroshi was able to secure funding from the Japanese government for both the ongoing habitat improvement at the original breeding site and the establishment of the new breeding site on Torishima Island.

The only other place where short-tailed albatrosses are known to have a nesting colony is on an island located southwest of Torishima. Hiroshi managed to visit this colony in 2001, but because the ownership of these islands is disputed among Japan, China, and Taiwan, it was extremely hard to get access.

A Very Patient Bird

There is also a place within US jurisdiction, the Midway Atoll, where short-tailed albatrosses have attempted to breed—although without success. No more than two individuals have been seen on any one of the atoll's islands at the same time, only one egg was laid, and there is no record of a hatching! Perhaps these stray short-tailed albatrosses are attracted by the sight or sound of the two million or so black-footed and laysan albatrosses that breed on those islands.

Judy Jacobs, who heads up the US Fish and Wildlife Service recovery plan for the short-tailed albatross, told me that one of these stray birds, believed to be a male, "has shown up on Midway's Eastern Island almost every breeding season since 1999." In 2000, to encourage a mate to join him, a number of decoys were placed on his island, along with a sound system playing recorded calls from Torishima. But despite these attractions, no other short-tailed albatross appeared, and year after year he waited in vain. Then his luck changed. "This year, just two weeks ago," Judy wrote in January 2008, "he was joined for the first time by another of his kind—a juvenile." The patient albatross and his new juvenile companion showed preening and pair-bonding behavior. "So perhaps," said Judy, "the adult bird's patience of nine years will finally be rewarded!!" I am longing to find out!

A New Island Home

The most important part of the recovery plan drawn up in 2005 by the US Fish and Wildlife Service, in cooperation with Japanese and Australian scientists, was to establish a new breeding colony in a safe place. In 2002, Torishima volcano had erupted again (it is one of the most active in the area), and although on that occasion it just spewed out ash and smoke—at a time when all albatrosses were out at sea—it was a stark reminder of the danger faced by the still-precarious short-tailed albatross population. It was important to try to establish a new colony on an island that was safe from volcanic activity and one that was accessible for monitoring. After much discussion, and a reconnaissance trip by Japanese scientists, Mukojima Island, one of the Ogasawara Islands about two hundred miles south of Torishima, was selected as a site for the new colony. Short-tailed albatrosses had been recorded breeding there as recently as the 1920s.

Before attempting to translocate precious short-tailed albatross chicks to Mukojima Island, a Japanese team of biologists from Yamashina Institute decided to work out albatross-chick-raising techniques with the non-endangered black-footed albatross species. This exercise was not very successful, but valuable lessons were learned that led to the development of better rearing techniques. So that the following year, when ten non-endangered black-footed albatross chicks were translocated to a specially prepared site on Mukojima Island, all but one of them fledged.

This success gave all those involved the courage to translocate the first precious short-tailed albatross chicks to Mukojima Island. There was a great deal of publicity in anticipation of this event. Fortunately, Judy Jacobs wrote me, things could scarcely have gone better. Ten chicks were transported from Torishima to their new home by helicopter in February 2008. And to the huge relief of everyone, all ten fledged—just a bit earlier than their peers on Torishima Island.

Today new technology is enabling scientists to find out exactly where the young short-tailed albatrosses spend their four to five years at sea after fledging. Twenty young albatrosses were fitted with tracking devices. Some of them flew straight from Torishima to the Bering Sea, traveling some four thousand miles in one month. This is an extraordinary journey, undertaken with no parental guidance, since the adults leave the breeding ground several weeks before the young. Of course it was particularly important to keep track of the birds that fledged from Mukojima. Five of them were equipped with satellite transmitters, as

Adult short-tailed albatross about to land on Torishima Island. Amazingly, 231 chicks fledged during the 2006–2007 breeding season and the population of the main colony was up to almost 2,000. *(Hiroshi Hasegawa)*

were five from Torishima. In September 2008, I got an update from Judy: All ten, she said, "are now foraging—and doing whatever else young albatrosses do—off the Aleutian Islands in Alaska." Five from Torishima and five from Mukojima!

The recovery plan for the short-tailed albatross, Judy told me, calls for translocations to Mukojima to continue for four more years, in the hope that by the fifth year some of the 2008 fledglings will return to Mukojima as breeding birds. And it is hoped that the decoys and sound system on the island may attract others of the species to also nest there. "It's a lot of work," Judy told me, "but very satisfying to play a part in the restoration of this magnificent seabird."

The "Patron Saint" of the Short-Tailed Albatross

I asked Hiroshi how he felt now that other scientists were actively involved in short-tailed albatross protection. "It makes me very happy," he said, "that conservation work that I initiated alone by myself more than thirty years ago has now developed into an international joint project to form a new colony." He will continue to monitor the situation on Torishima Island, and ensure that there are chicks to be translocated to

Hiroshi Hasegawa has devoted the past thirty-five years of his life—risking life, limb, and terrible sea sickness—to restoring this glorious seabird. Shown here, standing at the edge of Tsubame-zaki cliff, on Torishima Island, where he has just finished counting the short-tailed albatrosses (the tiny white dots clustered on the right, near the water) in the nesting slope below. *(Hiroshi Hasegawa)*

Mukojima. He has also set up the Short-Tailed Albatross Fund to receive contributions from the public. (You will find out more about this fund in "What You Can Do" at the end of this book.)

After working with these magnificent birds for so long, I wondered whether he had ever had a special relationship with any particular albatross. Not really, it seems, but there is the special pair that first nested at the new site he chose on Torishima in 1995. For twelve years now, they have maintained their bond, returning every year to the identical place to raise their chick. "And I will keep watching them," Hiroshi told me. His eyes lit up and for a moment he seemed far away, back in spirit in the wild places with the birds that, but for his efforts, might be no more.

THANE'S FIELD NOTES

Blue-and-Gold Macaw
(Ara ararauna)

When I first went to Trinidad with my colleague Bernadette Plair, I was treated to a remarkable journey that was at times hot, buggy, sleepless, bat-infested, and Spartan-like. Journeys are often defined by what you do *not* have, punctuated by unexpected gifts unavailable in your normal day-to-day. What I experienced on this trip was the opportunity to see more than a hundred species of birds in just two weeks, the most notable of which was the reestablished blue-and-gold macaw, a brightly colored and loud bird near and dear to Bernadette's heart.

Bernadette was born in Trinidad and grew up in the Sangre Grande area of the island. A soft-spoken woman with innate island diplomacy and keen tenacity, she has played a pivotal role in the conservation of her native wildlife. Like many "Trinis," Bernadette is of African, French, and East Indian descent, and recalls as a youngster in the 1950s and 1960s seeing and hearing the blue-and-gold macaws that the island was once famous for. "When I was a little girl," she told me, "I would see these beautiful and brightly colored birds flying above the canopy of palm trees, and naturally I never imagined that they could ever disappear."

These raucous birds are hard not to notice. Macaws are the largest and among the loudest of parrot species—and the blue-and-gold macaws are particularly striking with their vibrant royal blue wings and tail, which frame their nearly electric golden-yellow breasts. Unfortunately, the bird is particularly popular as a pet, and by the early 1960s it was extirpated from the island.

Their disappearance from Trinidad was actually the result of a number of factors. Illegal rice farming in the Nariva Swamp area of East Trinidad altered the bird's habitat. Blue-and-gold macaws rely on the palm trees on the edges of the swamp to build their cavity nests, and as the trees fell, so did the numbers of birds. Poachers cut down the hollow palms to raid the nests of young chicks and export them for the pet trade. Although illegal, and often controlled by the same people who traffic in illegal drugs, the shipping of parrots from throughout much of the tropics continues today.

Bernadette now lives in Cincinnati, Ohio, and is a research scientist at the Cincinnati Zoo and Botanical Garden's Center for Conservation and Research of Endangered Wildlife. During her twenty years at CREW, she has worked with many endangered species, from collecting data on the growth rate of the first captive-born Sumatran rhinoceros calf in 112 years to cloning endangered tropical plant species. All the while, she would visit her native home once each year to be with family, often noticing that most of the same problems for the island's wildlife still existed.

Poaching remained abundant, there was a lack of game wardens, and habitat loss due to illegal farming and development was growing. "These problems appeared to be getting worse each time I would go home," she said, "and it was a great concern because I could tell, objectively, what was being lost."

Rather than wait for others to do something, Bernadette decided to found CRESTT, the Centre for the Rescue of Endangered Species of Trinidad and Tobago. Initially, her idea was to start with what seemed a relatively simple project—bringing the blue-and-gold macaw back to Trinidad. After all, their historic range of Nariva Swamp was designated a protected 15,440-acre wetland in 1993. Bernadette's hope was that with this new protected status, putting birds back in the area would be a relatively quick and easy accomplishment. "Our hopes were indeed high in those early days," she said to me.

However, initial attempts to launch the program with confiscated birds met with no success. These adult birds, rescued from the pet trade, were not willing to breed in captivity on the island. They also suffered from the typical handicaps of captive animals reintroduced into the wild. The rescued macaws were naive to predators and vulnerable to new diseases, and had difficulty thriving. Still, Bernadette didn't lose hope. In fact, CRESTT continued to gain momentum. Bernadette garnered greater support from Trinidad's Wildlife Section and Forestry Division, as well as from international NGOs, including the Endangered Parrot Trust, Florida Avian Advisors, and the Association of Zoos and Aquariums.

And by 1999, an effective pilot project was under way: Eighteen young parrots were collected in Guyana by a licensed trader, in the hope that they would eventually form nine breeding pairs. The birds were transported from the forests of Guyana into special pre-release cages in Nariva, where they could become acclimatized to the surrounding trees and swamp.

This new system of translocating birds worked better than using captive-bred macaws. The Guyanan macaws brought with them their natural experience and savvy for survival in the wild. They quickly filled the niche vacated forty years ago in the Nariva Swamp and soon took hold.

With these releases came success at last, but also more work for Bernadette and her CRESTT team. As everywhere, it takes a multipronged approach for conservation to work in Trinidad. As Bernadette knows, "Conservation is never completely done. The work goes onward and onward." Government officials had to stay informed and involved in order to keep game wardens in the protected Bush Wildlife Sanctuary region of the swamp. Teams of volunteers had to be rallied to feed and water the birds while they were in big pre-release cages in the swamp. And small groups needed to camp at night near the birds to ensure their safety from wild or potentially even human predators—a very buggy but rewarding experience.

Public education was essential to long-term success, so as to eliminate interest in again taking the macaws out of nature. Everything from newspaper coverage to television stories and a billboard campaign declaring WELCOME HOME! to the beautiful parrots made sure not a single "Trini," or native to the island, could fail to know of the return of this once vanished creature. The result is that the blue-and-gold macaw is *the* flagship species of conservation in Trinidad. It is a source of pride symbolizing both the beauty of the island and the islanders' tenacity in bringing the birds back from the brink of extinction.

Perhaps the most joyful part of the ongoing efforts is how the Nariva Swamp and the macaws in particular have been embraced by many schools in Trinidad. Colorful festivals, parades, and musicals are all regularly performed by schoolchildren celebrating the natural heritage of Trinidad and how, with care, there is room for nature and people.

Today, a decade and a half after Bernadette's initial setbacks, the birds are solving their own problems. Nine of the initial parrots survived, several living in breeding pairs. In 2003, another seventeen wild birds from Guyana were released to provide new genetic stock.

To date, twenty-six of the thirty-one released birds have survived, and thirty-three chicks have been produced since the first releases in 1999. And any birders worth their salt will see macaws flying over the Nariva Swamp if they spend even a day in the area. But as much as the beautiful macaws, the children are what gives Bernadette hope. "I truly love seeing these young Trinis," she told me, smiling, "who—just like I did fifty years ago—stop on their way home to point and marvel at such a beautiful sight as a flock of macaws."

PART 5

The Thrill of Discovery

Introduction

As a child, I longed to be an intrepid naturalist, setting off into the unknown to discover new lands, and especially new kinds of animals. I think all children are born with a desire to discover things for themselves. They are curious, they want to investigate and learn about their exciting new (to them) world. And in the course of this, they make wondrous personal discoveries.

I was as exhilarated as any explorer in olden days when my friend and I crept out at midnight on a forbidden trip to a small, wild, undeveloped plot and discovered, in the moonlight, that a pair of barn owls had their nest there. It was a real adventure, for they swooped down and threatened us fiercely when we got too close—something I think back on when I read about all those who risk the wrath of the adults when they clamber up dangerous cliffs to inspect the eyries of birds of prey. That plot is built over now, the barn owls long gone, driven out by relentless development of the wild places.

I have been fortunate in my life—born in time to see some of those wild places before they were spoiled. And I treasure the memories of how things were. But there is still much to discover. Just yesterday (August 2008) came the report from Central Africa of lowland gorillas found in large numbers—doubling the estimated number of this endangered species. When I heard about those gorillas, it took me back to the few days in 2002 that I spent with Mike Fay and Michael "Nick" Nichols in the ancient, never-logged forest of the Goualougo Triangle in the heart of Congo-Brazzaville. When they first went there, they found animals that had never learned to fear humans—for even the pygmy hunters had not crossed the great swamps that protected the area for so long. Indeed, those swamps would have deterred anyone except Mike—but he found a secret way through and invited me for a visit. The journey started in a truck along a disused logging trail. Then came an enchanted time of silently moving along a gentle river, poled in our piraques by pygmy guides. And then a very, very long walk.

When at last we reached the camp in the forest it was after 10 PM, and I was too tired to appreciate anything—except the campfire and a deliciously simple meal cooked by the pygmies. But the next day, as I walked under the tall and ancient trees, I thrilled to the magic of a place that had not been explored by humans—at least not for hundreds of years. I put my hand on the trunk of one of those forest giants, sensed the rising sap, and knew great joy because, thanks to Mike, that whole forest is now a protected area. Safe—for the gorillas and chimpanzees and elephants. And for the trees. Because of Mike and others who care, many forests in Gabon have also been listed as protected.

In 2006, there was an expedition to the wild "Heart of Burma" where many new or thought-to-be-extinct species were found. Even more recently, an expedition to the remote Yariguies Mountains of Colombia discovered a fascinating array of species new to science. As did another to the wild, remote wilderness of the Foja Mountains of Papua. One benefit from these expeditions is that by discovering and writing about the last of nature's undiscovered wilderness areas, it is usually possible to get local and international support and pressure to protect them for future generations.

In the three chapters of this section, we share stories of discovery. Some of the discoveries are exotic—a new kind of monkey, a cave system sealed off from the outside world for at least five million years, a fish known only from fossils unearthed from the Devonian period—*sixty million years ago*! These are the stories that capture the imagination of the general public, creating headlines in international newspapers. Other discoveries seem less exciting, and are heralded simply by short notes in the local press or some specialist journal. Yet they are often thrilling to the biologists who find them—I have spoken to several, and their enthusiasm is contagious, shining from their eyes or sounding in their voice as we talk over the phone.

It is not just the joy of discovery—it is knowing that the life-form is important in the scheme of things. It all depends on your perspective. After all, it will make little difference to an elephant if a small plant vanishes; it will make all the difference between survival and extinction to a butterfly whose larvae feed exclusively on the leaves of that plant. And the biologist knows that all living things are interconnected in the web of life; that losing even the smallest strand can have unforeseen consequences.

It is true that we are experiencing the "sixth great extinction on earth," with thousands of species (mostly small, endemic invertebrates and plants) disappearing, forever, every year. And while we sink into despair or anger as we see how our own prolific and self-centered species continues to destroy, there is yet this feeling of hope. There are surely plants and animals living in the remote places, beyond our current knowledge. There are discoveries yet to be made. And the stories we share here, reports of fascinating new species discovered or rediscovered, give me new strength to face and fight the challenges that threaten our still-mysterious, still-magical planet.

This cavern and lake existed unknown and unseen by humans for about five million years. Israel Naaman was one of the first people to enter this cave and discover its secret. He took this photo of his friend Eitan Orel, who helped him map the cave. *(Israel Naaman)*

New Discoveries:
Species Still Being Discovered

So many of the books I read as a child were about intrepid explorers setting off into the unknown. They faced danger and tough conditions—and they came back with tales of strange and often fearsome creatures, then quite unknown to the Western world. It was hard to separate truth from fiction. There were descriptions of terrifying tribes fiercely attacking white strangers with spears; of cannibals with pointed teeth; of strange hairy creatures, half human and half animal, living deep in the forest. There were terrifying sea monsters that could sink a ship and mermaids luring sailors to a watery death. Gradually myth gave way to fact. The hairy men revealed themselves as great apes, the sea monsters were probably giant squid, and the mermaids were probably sea cows—dugongs or manatees. Linnaeus worked on his classification of the families, genera, species, and subspecies, arranging the animal and plant kingdoms into neat order. Charles Darwin sorted out how they got to be the way they were.

Gradually, during the past fifty years or so, discoveries of new species among the larger mammals and birds have become less and less frequent. But they have not stopped. And for those scientists studying the invertebrate hordes, finding a new species is, for the most part, no big deal—although, as we shall see, there are some pretty exciting finds in this area also. New fish and amphibian species are discovered quite frequently and, as we shall see in this chapter, there are occasional thrilling descriptions of larger creatures found.

I find it incredibly inspirational that even now, near the end of the first decade of a new century, with our planet groaning under the explosion of human populations, with the natural world retreating every day before the onslaught of development, there are still places where countless small creatures are living unseen by prying scientific eyes. This is so even in the developed world, but they are mostly found in remote, hard-to-reach rivers and lakes, mountainous forests, hidden caves, and canyons deep in the ocean. And then, during some expedition, they are spotted, their secret lives revealed. Sometimes the area is so remote, so undisturbed, that even larger birds and mammals can be found as well.

How thrilling to discover something that has never been described—probably the dream of every biologist who ventures into new terrain. When I arrived in Gombe in 1960, it was a very remote place. Apart from a couple of game wardens, few white people had ever been there. And many a time, as I gazed at some brilliant beetle or fly, or found a tiny fish high up near the waterfalls of the small swift streams, I wondered whether, perhaps, I was looking at a species unknown to science. Almost certainly, sometimes I was. For scientists working with plants, invertebrates, and fish are constantly identifying new species, especially now that DNA research enables us to make more rigorous distinctions between similar organisms.

In this chapter, I have selected a few of the discoveries made since the turn of the millennium, including previously undescribed birds and monkeys. They are not, for the most part, new to the people living there, who usually have names for them. But they are new to science, and for those who make such finds this is exciting, as each one adds to our knowledge of life on earth. There is just one problem: When a new species or subspecies is discovered, it has long been held that it can only be described, as for plant species, from so-called type specimens. Which means killing a few of the new creatures and putting their skins or whole bodies in preservative.

In the days when I worked for Louis Leakey at the National Museum (then the Coryndon Museum) in Nairobi, it sickened me to see drawer upon drawer of dead animals—the type specimens of not only invertebrates but also fish, amphibians, reptiles, birds, and small and medium-size mammals—and often there would be many of each. In addition, there were all those that had been skinned, stuffed, and put on display—and these of course included lions, chimpanzees, and so on. Such collections in museums around the world represent killings on a massive scale. Indeed, Dr. Thomas Donegan maintains that the killing of individuals for type specimens and for museum displays may have actually contributed to bird extinctions. In 1900, for instance, Beck collected nine of the only eleven individuals he had observed of a large and very rare bird, *Polyborus lutosus*, endemic to a small island off the coast of Mexico. Since then, this bird has never again been seen in the wild.

To Kill or Not to Kill . . .

Today, as we face mass extinctions on our planet, more and more scientists believe that it is ethically wrong to kill newly discovered creatures that are rare and most likely endangered, and that new technologies mean that it is not *necessary* to obtain dead specimens. This has led to a heated and sometimes acrimonious ongoing debate. For example, Alain Dubois and André Nemésio describe those who are against killing for science as an "ethically correct tyranny" who peddle "a hypocrisy and a lie" and choose "ignorance in the name of conservation." Donegan counters that the International Code for Zoological Nomenclature defines the term *specimen* as: "An example of an animal, or a fossil or work of an animal, *or of a part of these*" (my italics). Thus, argues Donegan, it is possible to achieve one's objective using nonlethal methods, describing a new species through meticulous descriptions and photographs, along with hair or feather samples and blood for DNA analysis.

Drs. Dubois and Nemésio also believe that if a newly discovered species is known from just one individual, it is probably as good as extinct anyway, so it may be better to kill it for a type specimen rather than risk that it disappear unrecorded by science. But suppose, says Donegan, another individual is subsequently found? In part 4, we describe how the black robin population bounced back from a low of just one remaining female and four males.

While the scientific debate continues, it is comforting to know that a growing number of previously undescribed species have been documented without using dead specimens—and that the descriptions have been generally accepted, and published in peer review scientific journals.

Donegan makes another important point: Researchers who seek to convince poor rural communities that scientific collecting is justified, while hunting or animal trade should be controlled or prohibited, are likely to be regarded as inconsistent and are setting a terrible example. Those who describe species without killing them have the moral authority to encourage conservation initiatives among the local people—

in whose hands the future lies. When JGI was working in Burundi, I decided to end a collaborative arrangement with another organization when I found it was planning a large-scale collection of birds and small mammals for scientific research in our study area. I pointed out that we had spent a great deal of time convincing the local population that wildlife should be respected and protected and that if they were now offered money to go trap and kill them, all our headway would be lost.

New Primates—Our Closest Relatives

Two new species of Old World monkeys—in the Himalayas and in Tanzania—and one New World monkey in Brazil, have been found since the start of the new millennium. In 2003, the Nature Conservation Foundation organized an expedition to the mountainous Indian state of Arunachal Pradesh, bordering Tibet and Myanmar. They found a monkey unknown to science—the first macaque species to be discovered since 1908. Of course the local people knew the animals well and called them *mun zala*—the "deep forest monkey"—which led to its scientific name of *Macaca munzala*, commonly known as the Arunachal macaque or stocky monkey. Fourteen troupes of about ten monkeys each were located in areas of undisturbed forest—the monkeys were shy and very wary of people. They are, as one of their names suggests, stocky in shape, with brown fur that is darker on their heads, and short tails.

Our second monkey, *Rungwecebus kipunji* or the kipunji, was found in 2003 in the southern highlands of Tanzania. By an almost unbelievable coincidence, it was discovered in two different locations some 250 miles apart, at almost the same time, by two completely separate expeditions! Dr. Tim Davenport of the Wildlife Conservation Society (WCS) and his team first found the kipunji in the Rungwe-Livingstone Forest in December 2003.

Less than a year later, in July 2004, Dr. Trevor Jones led an expedition sponsored by the University of Georgia into the Ndunduhi Forest Reserve of the Udzungwa Mountains and discovered four groups (each group has about thirty to thirty-six individuals) of kipunji living there as well. Sadly, I was told by Tim Davenport that this kipunji population is no longer considered viable, despite the fact that the reserve has been highly protected.

Meanwhile, the Mount Rungwe–Livingstone Forest has been heavily logged and there have been many poachers. Even so, Tim Davenport's team has since discovered as many as thirty-four groups of kipunji living there—bringing the total number of individuals up to 1,117 as of March 2009. Fortunately, the Mount Rungwe–Livingstone Forest is about to become a nature reserve (something Tim and his team have fought hard for), which should help keep the kipunji more secure.

The really exciting thing about this discovery is that the monkey is not merely a new species, but a completely new genus, having biological characteristics that differentiate it from both mangabey and baboons. (For those who don't remember their school biology lessons, genus is an even broader classification than species.) At first it was thought to be a kind of mangabey and named the highland mangabey, but then a dead one was found, trapped by a local farmer, and DNA analysis showed that it was more like a baboon. It is about three feet in length, with long brownish fur, a crest of hair on its head, and pronounced whiskers on its cheeks. Instead of communicating with a mangabey-type *whoop gobble*, the kipunji has a *honk bark*. Just reading about these sounds makes me really want to hear them for myself. It certainly stimulates the auditory imagination—a *whoop gobble* and a *honk bark*.

In an interview, Dr. Jones said, "I'll never forget the day we were surveying biodiversity in the forest, and one of our team suddenly grabbed me and pointed to a monkey in a tree a hundred meters away. I grabbed my binoculars and nearly fell over. It was a very surreal moment, and I simply stood there in disbelief." Soon after this fantastic experience—surely every biologist's dream—he learned about the new monkey that Davenport and his team had just found. When they subsequently realized that the two new monkeys were the same species, they decided to publish their findings jointly. It has long been known that the mountains of southern Tanzania have provided a refuge for a variety of species long extinct elsewhere—what else, I wonder, is waiting to be found?

The New World monkey, the blond capuchin (*Cebus queirozi*), was discovered near Rio de Janeiro, Brazil, in 2006 by Antonio Rossano Mendes Pontes. It has golden hair with a white "tiara" on its head. Thirty-two individuals were seen in a forest and swampland fragment of only about five hundred acres. One individual was caught, examined, photographed, and returned to the forest. Some suspect that rather than a new species, the blond capuchin may be a rediscovery of a monkey named *Simia flavia*, known only from a drawing by German taxonomist Johann Christian Daniel von Schreber in the 1770s.

Primates from Brazil and Madagascar

In the vast forests of the Amazon Basin, many secrets of nature still lurk. My longtime friend Dr. Russ Mittemeier, who now holds the prestigious position of scientific director of Conservation International, has spent many years exploring the Brazilian Amazon forests. Between 1992 and 2008, he and his team discovered, described, and named a total of six new marmoset species and two species of titi monkeys. One of them, for me, was very special because on a brief visit with Russ I was able to meet the little creature. Russ had only recently rescued her from a remote village. Diminutive and absolutely enchanting, this scrap of a primate sat on Russ's shoulders as he told me stories of his travels.

Presently she moved onto my shoulder, and I had an unreal feeling—I was in contact with a tiny being that only a handful of Westerners had yet seen. How many of her kind, I wondered, were out there, living their unknown lives? It was subsequently determined that she represented a completely new genus. Now known as *Callithrix humilis*, the black-crowned dwarf marmoset, she has a name longer than she is! In fact, during the first eight years of the new millennium, a total of eight new species of prosimians (all primates other than monkeys and apes) have been described in Brazil: three marmosets, three titi, and two uakari.

During the same eight years, no fewer than twenty-two new species of lemurs were discovered in Madagascar—seven species of the tiny mouse lemur, two giant mouse lemurs, five dwarf lemurs, two woolly lemurs, and four new sportive lemurs. Russ has also spent time in Madagascar, and in 2006 a mouse lemur and a sportive lemur were both named for him.

New Birds

Whenever a new species of bird is found, a ripple of excitement runs through the circles of the ever-growing bird-loving public. In 2007, Dr. Blanca Huertas, of the Natural History Museum of London, led an expedition to the remote Yariguies Mountains in Colombia, and among many fascinating discoveries that she and her colleague, Thomas Donegan, made was the Yariguies brush finch (*Atlapetes latinuchus yariguierum*), a small bird with striking black, yellow, and red plumage. I spoke briefly on the telephone with Blanca and asked her how she felt when they found this bird.

"It took time before it sank in," she said. She thought for a moment, then added, "I think it is wonderful that we have left a little fingerprint on science." (Her team left another little fingerprint when they found a new species of butterfly.) Blanca told me that this brush finch was the first bird species from the New World in which individuals were not killed deliberately to provide type specimens. Instead the team planned to identify the species by means of detailed descriptions, photographs, and blood samples. In fact, one of the two birds caught for this purpose died accidentally, so they ended up with a dead type specimen after all.

For the past few years, environmentalists have been pushing to have the area protected; the finding of the new species has been of great help in this respect. Blanca told me that the area will very soon be declared a national park.

Ant from Mars

In mid-2008, short articles appeared in many international newspapers about a newly discovered ant from Brazil's rain forest—the Ant from Mars. Soon after I saw one article, I tracked down Christian Rabeling, the biologist who discovered it near Manaus, and we had a fascinating conversation. The most exciting thing is that this ant is not just a new species, but a new genus. Its closest relatives seem to be ants that lived some ninety million years ago. I asked Christian how he felt about this discovery: "I think someone really loves me!" he said.

He found the pale, eyeless ant by pure chance. One evening, when it was nearly dark, he was sitting in the forest getting ready to go home. He saw a strange white ant walking over the leaf litter and, not recognizing it, popped it into preservative in one of the small vials that he always carried and put it into his pocket. When he got back home, he was tired and had quite forgotten about it. Three days later, he found the specimen in the pocket of his pants. It was then that he realized he had found something extraordinary. Subsequently he sent photos of his specimen to Stefan Cover, who is in charge of the ant collection in the Museum of Comparative Zoology, the largest ant collection in the world.

Stefan told us his reaction. "The first glimpse I got was just a grainy image on Christian's computer. But it was obvious right away that I was looking at a completely unusual animal. I said, 'Holy smoke. I don't know what the hell that is.'" Later he told us, "Usually I know an ant when I see one. I can identify the subfamily, what genus, in some cases I can even guess the species. But looking at this ant made my brain go into tilt. It was definitely an ant—but unlike anything I had ever seen."

Stefan, knowing that Ed (E.O.) Wilson, famous among other things for his definitive book about ants (and a great hero for Christian), would love to see the strange ant, fetched him from down the hall. And Ed, looking at the image on Christian's computer, made his now famous comment:

"My God! That looks like an ant from Mars."

"It was thrilling for all of us," said Stefan. "A lot of scientists live for this moment."

There is a final quirk to this story. Five years before Christian made his find, Manfred Verhaagh had found two strange-looking ants in just one of several soil samples from the area where Christian was working. Manfred preserved them in a vial—but as he was traveling with them, taking them to be identified, the container leaked and the priceless specimens were utterly destroyed. They tried everything to rehydrate them, but nothing worked. When five years later, Christian found the "Ant from Mars," he sent a photo to Manfred—who immediately knew it was the same as the two that had been destroyed!

Science Fiction in the Depths of Sea and Earth

As we've mentioned, new species of invertebrates are continually being found. But sometimes a discovery seems out of the ordinary, especially when we find a survivor from a world millions of years ago, when life-forms were struggling to survive in the inhospitable environment of the cooling planet.

Such is the case with the recent discovery of giant tube worms in the depths of the Gulf of Mexico by marine biologists from Penn State University. In this unlikely and eerie world, the worms live on chemicals from volcanic vents on the ocean floor. They have no natural predators and can grow to ten feet in length! The biologists, who measured the growth rate of individual tube worms over a four-year period, calculated that they would have to live for 250 years—a quarter of a millennium—to reach their maximum length. If this is true—if there were no growth spurts caused by changes in sea chemicals due to volcanic action—they would be the longest-living invertebrates on earth. Or at least of those that we have discovered—who knows what other wonders are out there!

The next tale is even more extraordinary—the discovery of the Ayalon

Cave (as it is now known) near Ramla in central Israel. The entrance to an extraordinary world was discovered accidentally when workers in a deep limestone quarry broke through the wall. When scientists from Jerusalem's Hebrew University got there, they found a whole unique ecosystem a hundred yards under the ground.

I managed to track down Professor Amos Frumkin of the Hebrew University, and he suggested that I contact his student Israel Naaman, who was one of the first to enter the cave. Israel described it as being a very big "maze cavern." The conditions, he said, were "not friendly for us—narrow passages, heat, and extremely high humidity." But, he went on, "the feeling of walking into an unknown place that no one has been before is incredible."

It is best if I quote Israel's words, for they give a real sense of the excitement that he and other members of the group felt at the time. "We arrived to a great round hall, forty meters diameter, with twenty-seven-meter height ceiling. I couldn't see the other side of the hall; darkness swallowed the headlight beam. I took out a stronger hand light and an amazing spectacle was discovered to me—a beautiful underground blue pond. The water was still and the other guy bent forward to the water and started screaming, 'There are animals in the water!' On the water surface there was a thin bacterial mat and in the water, pale crustaceans swam, up to five centimeters long and of lobster-like shape.

"Later, guided and equipped by biologists, we found in this lake and its surroundings, a very rich and vital ecosystem, including six

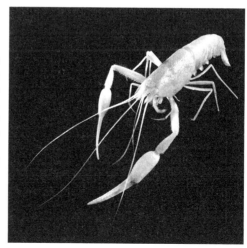

Soon after the team entered the cave, they discovered eyeless white creatures living in the subterranean aquifer. Ultimately, the team discovered six new species of arthropods. This one was named *Typhlocaris ayyaloni*. (Dr. David Darom)

new species of arthropod, four of them aquatic and two terrestrial. Additionally we found remains of two other species that probably had become extinct due to intensive pumping of water from the aquifer."

Subsequent DNA testing showed that all eight of the species found by Israel and the team—white, shrimp-like crustaceans and scorpion-like invertebrates, all of them without eyes and apparently feeding on the surface bacteria—were new to science. They were, said Professor Frumkin, "absolutely unique in the world."

Further exploration revealed a maze of passages extending for more than a mile, sealed from the surface water and nutrients above by a layer of chalk and drawing water from deep underground. The whole unique ecosystem dates back five million years, to when part of Israel was under the Mediterranean, and it has been closed off ever since. Unfortunately, as Israel noted, the underground lake is part of an aquifer that is one of the most important freshwater sources for Israel. This means that the cave and its whole ecosystem is affected and extremely endangered.

Let us be thankful we have at least learned of its existence and can marvel at the diversity of life-forms on our amazing planet. How easily it could have vanished, without due reverence, to join the other extinct life-forms of its mysterious prehistoric era.

Unexplored Forests of Indonesia's Foja Mountains

Some people find it hard to believe that there are still huge tracts of remote forests that have remained unknown to the outside world. During a recent visit to John Conaghan, my dentist in Washington, DC, I was telling him—during the odd moments when my mouth was not full of instruments and fingers—about this book. He told me that his neighbor, Bruce Beehler, had recently returned from an exciting expedition to Indonesian New Guinea. "He found some kind of new bird," John told me, and he gave me Bruce's phone number.

Bruce is an ornithologist, an authority on New Guinea birds as well as a tropical ecologist, and currently serves as a senior research scientist at Conservation International in DC. When we talked, he told me something about the expedition he had led, and gave me a link to his Web site. There I learned that the isolated Foja Mountains of Papua, Indonesia's easternmost and least explored province, lie on the western side of the great tropical island of New Guinea and probably represent the most pristine natural ecosystem in the entire Asia-Pacific region. The range comprises more than three hundred thousand hectares (some 740,000 acres) of old-growth tropical humid forest. The customary landowners of the Foja Mountains, the Kwerba and Papasena peoples, total only a few hundred individuals. They hunt and collect herbs and medicines from the fringes of the forest but seldom penetrate more than a few miles into the interior. With the human population so small, animals are still abundant within a mile or so of the villages, and the hunters have no need to travel farther afield.

The story leading up to and culminating with the expedition Bruce led is almost like a fairy tale. "For decades," Bruce told me, "the Foja Mountains had been a promised land to biologists in search of the unknown." In 1981, Professor Jared Diamond had managed to make two brief visits to the mountains and found what at least a dozen expeditions had failed to do—the home of the almost mythical golden-fronted bowerbird. It had been described by a German zoologist in 1895 from "trade skins" that had been collected in some unknown corner of western New Guinea, and despite at least a dozen expeditions sent out to search for its homeland, the bird had never been seen alive by Western scientists—until Diamond's visit eighty-six years later.

This exciting news triggered renewed ambitions to explore the Foja Mountains. Conservation International together with the Biology Research Center of the Indonesian Institute of Sciences made plans to visit the area to learn more about its wildlife. It was not easy—it took ten years to get the permissions that were needed from four government departments and a number of provincial and local authorities. Not until November 2007 did Bruce finally set off with his fourteen-member team of Indonesian, American, and Australian scientists. They were dropped down by a helicopter and set up camp in a remote mist-shrouded world high up the mountains.

Even with expectations running high, none of the members of that expedition could have imagined that, within minutes of arrival, they would encounter a bizarre red-faced and fleshy-wattled honeyeater bird unlike anything in their field guide. With a thrill, Bruce told me, he suddenly realized that he was looking at a completely new species—the first new bird discovered on the island of New Guinea since 1951. For the time being they named it the wattled smoky honeyeater. (You can see photos of this bird on our Web site: janegoodallhopeforanimals.com.)

And then, just one day later, the team was amazed when a male and female Kleinschmidt's six-wired bird of paradise (*Parotia berlepschi*) came right into their camp, and the male, with his spectacular plumage, displayed on the ground to the female for more than five minutes—in full view. "We stood in awe as the male romped about in the saplings, flicking his wings and white flank plumes, whistling his sweet two-note song for the female," said Bruce. "I was too spellbound to get my camera out that first time."

They were the first Western scientists to see the birds alive, and they instantly realized that it was a full and distinct species, looking very different from the other members of this genus of bird of paradise. The team had discovered the unknown homeland of this remarkable bird and seen its spectacular displays within two days of arrival! I can only imagine the excitement in the air as they gathered for supper that evening. And it was not long before they located one of the yard-high constructions of carefully placed twigs that marked the "maypole" dance grounds of the golden-fronted bowerbird and made the first photographs of this species displaying at its bower. It turned out that the bird was common in the area.

The discoveries continued, day after day. Altogether forty species of mammals were recorded, including many that are rare in other parts of New Guinea, but common and unafraid in the Foja Mountains. The long-beaked echidna, a relative of the duck-billed platypus, looks a bit like an oversized hedgehog but with a long snout. This is the largest of the bizarre and primitive egg-laying mammals known as monotremes. A few of these rare beings were seen on three successive nights. Twice they allowed themselves to be picked up and carried into camp for study.

These strange creatures have never reproduced in captivity, and nothing is known about their natural behavior. Another highlight was the discovery of a population of the golden-mantled tree kangaroo (*Dendrolagus pulcherrimus*)—the first record of the species in Indonesia. It is an extraordinarily beautiful jungle-dwelling kangaroo that literally climbs into trees and is critically endangered. This is only the second site in the world where it is known to exist.

The Foja Mountains appear to be one of the richest sites for frogs in the Asia-Pacific region—the team found more than sixty species, at least twenty of which are new to science. These mountains are also a paradise for butterflies—more than a hundred and fifty species were found, four new to science. And of course the botanists found many remarkable and previously undescribed plant species, including a rhododendron growing high in the treetops with spectacular scented white flowers, and five new species of palm.

I cannot imagine anything more fantastic than being part of such an exciting expedition. It was just such an adventure that I had dreamed of as a child. I asked Bruce how he felt when he got there. How was it to wake up in paradise?

"I remember standing, at dawn, in a lovely little bog atop a flat ridge in the very center of the Foja Mountains," he told me. "A mighty black sicklebill quipped loudly to the south. A dozen other birdsongs floated overhead. The sky was a deep blue. I was in an Eden of sorts, one without the footprint of humankind, one left to the birds and marsupials . . . it was a sublime moment."

When I spoke with Bruce, he told me that in two days he would be setting off on another expedition to his Eden—and I was left with an unrealistic longing to be part of it.

A Monster Palm from Madagascar

My last story is about a giant palm found recently on Madagascar. I learned the story of this palm during a visit to Kew Botanical Gardens in 2008. John Sitch, who works with palms, was eager to tell me about this extraordinary discovery. He picked up one of the row of pots that sprouted young specimens of the plant, holding it almost reverently. He is not a demonstrative man, but the excitement was clear in his voice as he explained that this was a completely new species of fan palm, the largest ever found in Madagascar—the adult leaves have a sixteen-foot diameter. Apparently the full-grown palm is so massive that it can actually be seen on Google Earth!

I can just imagine the amazement of Xavier Metz, the French manager of a cashew plantation, when he and his family came upon this huge palm as they were exploring a remote area in the northwest of the country. He had never seen anything like it, and was sure it was a new species, so he took photos.

It was even more exciting than anyone had thought—not only an undescribed species, but actually the single species of a new genus. And this genus was from an evolutionary line that was not known to exist in Madagascar. It was named *Tahina spectabilis*—*tahina* is Malagasy for "to be protected or blessed" (the given name of Anne-Tahina, daughter of the discoverer), and *spectabilis* is Latin for "spectacular." An intensive survey showed that there was just one population of ninety-two individuals tucked away at the foot of a limestone outcrop.

This palm has the most extraordinary life cycle. When it is about fifty years old and has reached a height of nearly sixty feet "the stem tip starts to grow, and changes into a giant terminal inflorescence sprouting branches of hundreds of tiny flowers," John told me. These flowers ooze nectar, and are soon surrounded by birds and insects. It is a spectacular flowering, "and each flower, once pollinated, can become a fruit," said John. Then, once the fruits have ripened, the palm is utterly exhausted. The flowering and the fruiting are its swan song, and it collapses and dies.

About a thousand seeds from this palm were carefully collected and have been sent to Kew's Millennium Seed Bank in Sussex. Seeds have also been distributed to eleven botanical gardens around the world, so that the palm can be conserved in living collections—one of the goals of the seed bank. Because *Tahina* is limited to just the one area on the island, and because flowering and fruiting are such rare occurrences, conservation at the site will not be easy. However, the villagers have become involved. A village committee has been set up to patrol and protect the area. And some of the seeds have been sent to a specialist palm seed merchant in Germany so that he can raise and sell palms to create funds for village development as well as conservation of the palm.

I told John that I look forward to seeing *Tahina* in Kew's Palm House, a spectacular public exhibit of species from around the world. But alas, I shall not be alive when the first Kew plant is fifty years old and has its first burst of flowering!

The Lazarus Syndrome: Species Believed Extinct and Recently Found

It is not only the finding of a species new to science that is exciting. To discover a living individual of a species long thought to have been extinct and lost forever is, in many ways, even more rewarding. It gives us a little hope to know that some individuals of a species that, after much searching in the wild, has been officially listed as "extinct" might, just might, still be around. Because then we can give it another chance.

I visited Ghana soon after the Miss Waldron's red colobus was pronounced extinct and met a biologist who believed that a group of these monkeys still existed in a remote, swampy part of the country. Immediately I wanted to go and search for them. Of course, I could not go, and anyway it seems the rumors were probably just rumors. But I could imagine the thrill of announcing to the world that these monkeys were not extinct after all. I can so well understand why people obstinately continue searching for some animal or plant that they feel certain is out there somewhere—if only they could find it.

Recently, when I was in Australia, I met those who felt sure that the "extinct" Tasmanian wolf still exists. Indeed, I was given a book listing all the recorded "sightings" of the creature. And people who know Tasmania describe remote, hard-to-penetrate forests where, they say, this animal could still exist. I was given the cast of a paw print of one of the last known individuals, and as I look at it I think . . . perhaps, just perhaps, his great-grandchildren are hiding out there.

The rediscovery of species thought to be extinct is known as the Lazarus Syndrome. Unlike their namesake in the Bible, they have not, of course, been resurrected from the dead—they were there all the

Nicholas Carlile with Eve, a Lord Howe's Island phasmid. Nicholas was one of the first two people to lay eyes on one of these giant stick insects in 2001, after it was presumed extinct in the 1920s. *(Patrick Honan)*

time. Some of them, such as Lord Howe's giant stick insect, are exotic and capture the imagination of the general public, creating headlines in international newspapers.

Other discoveries seem less exciting, and are heralded simply by short notes in the local press or some specialist journal. Yet these seemingly less significant finds are just as meaningful in the general scheme of things, for all living things are interconnected, and, as we said, removing even the smallest strand might have unforeseen consequences.

This chapter contains stories about some invertebrates, birds, and mammals that have been rediscovered. Sometimes quite by chance. Sometimes as a result of determined searching over long periods. And while it is true that we are indeed facing the "sixth great extinction" with thousands of small, endemic invertebrates and plants rapidly disappearing, it is encouraging to know that a few species that are thought to be extinct may be, just may be out there waiting to be rediscovered—and given another chance.

These are stories of precious life-forms that have been written off, consigned to the legions of the extinct—but have refused to die. Stories to give us hope.

Lord Howe's Island Phasmid or Stick Insect
(Dryococelus australis)

In 2008, during my lecture tour in Australia, I met a very large, very black, and very friendly female Lord Howe's Island stick insect. She crawled from one of my hands to the other several times, and when I gave her the opportunity, she also crawled onto my head and face. The encounter sent shivers up my spine—knowing, as I did, the almost incredible story of how she came to be there. Let me share that story.

Lord Howe's Island, small and partly covered with lush forest, is about three hundred miles off the coast of New South Wales, Australia. It was the only known home of the Lord Howe's Island phasmid—or stick insect, or walking stick—a giant creature about the size of a large cigar, four or five inches long and half an inch wide. Once they were found throughout the forests of the island, and known by the locals as land lobsters.

But in 1918, black rats arrived on the island when a ship was wrecked. And, as always, these relentless colonists quickly adapted to their new environment. Unlike all other stick insects in Australia, this giant phasmid lacked wings. And so it was an easy—and probably delicious—prey.

At some point in the 1920s, the Lord Howe's Island phasmid was presumed extinct.

Then, in 1964, rock climbers found the dried-out remains of a giant stick insect on Ball's Pyramid, an eighteen-hundred-foot-tall spire of volcanic rock, fourteen miles from Lord Howe's Island. Five years later, other rock climbers found two other dried bodies incorporated into a bird's nest. This remote pinnacle, the haunt of countless seabirds, is almost entirely without vegetation. It seemed impossible that a very large, forest-loving vegetarian insect could be surviving in such a bleak environment. And so biologists ignored these reports until, in February 2001, a small group of people—Dr. David Priddel, the senior research scientist of the Department of Environment and Climate Change (New South Wales), his colleague Nicholas Carlile, and two other intrepid souls—decided to settle the matter once and for all, and set out on what they felt sure was a wild-goose chase.

A Perilous Journey

In February 2007, from my home in Bournemouth, I had a wonderful talk with Nicholas Carlile (whom I met the following year). He told me that it had been a potentially dangerous undertaking. The seas around Ball's Pyramid are rough, and the team of three men and one woman had to leap from their small boat onto the rocks. ("Swimming would have been much easier," Nicholas told me, "but there are too many sharks!") His description of the landing—the desperate leap for the rocks with the boat surging up and down—was hair-raising. But they all made it, put up a small camp, and set off to climb as far as Gannet Green, about five hundred feet up the spire of rock where the main vegetative patches clung to life.

They searched the place thoroughly but found nothing other than some big crickets, and eventually the heat and lack of water drove them back down. Then, in a crevice about 225 feet above the sea, they came upon another tiny patch of comparatively lush vegetation, dominated by a single melaleuca bush. A small water seepage allowed this tiny oasis of plants to maintain its precarious foothold. Here they found the fresh droppings of some large insect, but assumed it was a cricket.

Back in camp, over supper, they discussed the situation. David Priddel knew that the stick insects are nocturnal, and that the group would have a better chance of seeing them if they went back to that bush at night. But he knew he could not make the climb in the darkness and so was loath to suggest it. Nicholas had the same idea, though, and he and Dean Hiscox—a local ranger and an expert rock climber—volunteered to make the almost suicidal climb in the dark. They set out with head-lamps and one instant camera. "It gives me the wobblies just to think of it," Nicholas told me over the phone.

Finally they reached the vegetated area. "And there is this enormous shining, black-looking body spread out on the bush," said Nicholas. "I yelled out some kind of expletive. And the two of us began celebrating like kids, jumping like six-year-olds"—but, he assured me, jumping with great caution, since the ledge was only thirteen feet wide on a sixty-degree slope, and it would have been very easy to slip over the edge!

Almost at once they saw a second giant insect stretched out on the vegetation. Nicholas's excitement, as he talked to me six years later, was palpable. "It felt like stepping back into the Jurassic age, when insects ruled the world," he said. "It was one of those iconic moments that changed my life forever. We kept telling each other that no living person had ever seen one of these giant insects." They also found a youngster, a nymph. Nicholas took three photos and then they had to try to calm down before embarking on the highly dangerous nighttime descent.

When they got back to camp, the others were sleeping. "I crept up to Dave," said Nicholas, "put my lips to his ear, and whispered, 'We found a phasmid!' Soon everyone was awake!"

Early the next morning, the whole team climbed back up and made a thorough search. They found some more frass—apparently the proper terminology for insect poo!—and about thirty eggs in the soil. Then they had to leave as the boat was picking them up at 10 AM. The ocean swell had increased considerably by the time they left: The boat was rising and falling ten feet every few seconds. It meant split-second timing for the jump onto the deck—it gives *me* the wobblies just thinking about it!

They were all convinced that the only population of Lord Howe Island's giant phasmid in the world lived on that one shrub.

How did the little colony get to that isolated pillar of rock? Perhaps a gravid female had made the fourteen-mile journey from Lord Howe's Island clinging to the legs of some seabird, or to floating vegetation after a storm. And once there, how had she found the one and only suitable habitat on the entire pyramid? Perhaps, suggested Nicholas, a recently dead female containing eggs had been picked up as a "stick" on the main island and transported to a seabird's nest near the phasmids' bush. But however she got there, how on earth had her descendants survived for eighty years or so in that desolate environment? We shall never know.

As soon as they got back, the biologists set to work to draw up a draft recovery plan for the phasmid. They faced many battles with bureaucracy, and two years elapsed before they had permission to return—and they were only allowed to catch four individuals. They found that there had been a big rockslide on Ball's Pyramid. How easily the entire population could have been wiped out during those two frustrating years. However, on Valentine's Day in 2003, they found the colony still thriving on its one bush. To transport the incredibly rare cargo—the four captured insects—a special container had been prepared, and this presented a problem when they arrived in Australia. It was not long after 9/11 and security was very tight, yet they had to convince the officials not to open the precious box!

One of the scientists on that second expedition was Patrick Honan,

The team of discoverers approach the treacherous Ball's Pyramid, fourteen miles off Lord Howe's Island—a tiny population of phasmids mysteriously found its way here and lived unknown for eighty years. (Nicholas Carlile)

The team assists one another across a shore-line traverse—in search of the elusive Lord Howe's Island phasmid. The sea conditions had deteriorated overnight, and rising seas meant the team only had limited time on the Pyramid. *(Nicholas Carlile)*

a member of the Invertebrate Conservation Breeding Group (among many other things), who subsequently played a key role in the future of the phasmids. One pair went to a private breeder in Sydney, and the other two (Adam and Eve) went with Patrick to the Melbourne Zoo. To everyone's delight—and relief—Eve soon began laying pea-size eggs.

But within two weeks of captivity the pair in Sydney died, and Eve became very, very sick. Patrick worked every night for a month desperately trying to cure her. He scoured the Internet for help, but no one knew anything about the veterinary care of giant stick insects! Eventually, based on gut instinct, Patrick concocted a mixture that included calcium and nectar, and fed it to his patient, drop by drop as she lay curled up in his hand. To his joy she seemed to get better and laid eggs for a further eighteen months. But the only ones that hatched were the thirty or so that she'd laid before she fell sick. How fitting that the first of these hatched on International Threatened Species Day! I can well imagine the excitement and sheer delight of all those concerned when out crawled a bright green nymph—already almost an inch long.

It was in 2008, when I visited the Melbourne Zoo, that I met Patrick and he introduced me to that friendly female stick insect I described at the start of this story. She was, he told me, one of the fifth generation of these phasmids in captivity. Patrick showed me the rows of incubating eggs—11,376 at the last count, he said. And there are about seven hundred adults in the captive population. They are very special insect beings. Patrick showed me a photo of how they sleep at night, in pairs, the male with three of his legs protectively over the female beside him.

Then we went for a ribbon-cutting ceremony. Surrounded by the whole team, I wielded the scissors and declared that the zoo's brand-new Lord Howe's Island stick insect exhibit was now officially open. Later, Patrick told me he'd left academia, believing that the most important conservation is grassroots—that people will only try to save animals once they get to know them firsthand. He has just completed the final planning of a project to have these stick insects reared by a hundred primary and secondary schools—a fantastic opportunity for the students to become involved in an ongoing conservation program in their own classrooms.

As a further insurance for the species' survival, eggs are now being sent to other zoos and private breeders in Australia and overseas. The two hundred eggs that had been sent to the San Antonio Zoo in Texas have already begun to hatch, Patrick told me: "So the species has now gone international."

With so many of the giant insects thriving, there is an increasingly urgent need to release the species back into the wild on Lord Howe's Island. And this is giving a significant push to the program to eradicate rodents there planned for the winter of 2010. Once they are gone, the first giant phasmids will be returned to the place of their ancestors.

It has been an incredible story. Nicholas told me that when he joined David on that first expedition to Ball's Pyramid, they both believed it was doomed to failure. How could a creature, last seen eighty years before, possibly be alive on a piece of barren rock way out in the ocean?

"So," said Nicholas, "we went with the purpose of proving the phasmids *not* to be there, to refute, once and for all, on good scientific evidence, the rumors about their existence. Which just goes to show!"

The Mallorcan Midwife Toad *(Alytes muletensis)*

My childhood natural history bible, *The Miracle of Life*, described the fascinating life history of midwife toads. The female lays the eggs, but the male carries and protects them until they hatch. It was one more of those stories that left me increasingly fascinated by "the miracle of life."

There are five species of midwife toads, widespread across Europe and northwest Africa, but the existence of the toad on Mallorca, an island off the east coast of Spain, was not known until 1977 when fossilized remains were discovered there. At that time, it was thought that it had been extinct on the island for about two thousand years. And then, just three years later in 1980, one single individual was found in a deep canyon in a remote mountainous region in the north. This led to the discovery of a small population living there.

They are golden brown to olive green in color, with patterns of darker brown or black, and large eyes. Like most toads they are nocturnal, hiding under rocks during the day. The females produce strings of eggs, which the father fertilizes externally then wraps around his ankles. He then carries his cumbersome load of between seven to twelve eggs, making sure to keep them moist, until they are ready to hatch. At that point, he enters shallow water until all the—exceptionally large—tadpoles have emerged and swum away.

I learned about the program to save the remaining Mallorcan midwife toads firsthand from Quentin Bloxam, a scientist with the Jersey Wildlife Conservation Trust, who happened to be in Mallorca at the time of the discovery in 1980. "There was a student there at that time studying tortoises," Quentin told me over the phone, "and he came to discuss his project with me and ask my advice." At the end of this meeting, the student asked him: "By the way, have you heard about this toad that has just been discovered?" This was news—and exciting news—to Quentin, and he set off with the student down a small street to meet Dr. J. A. Alcover, the biologist who had made the fantastic discovery. "We went into his office," said Quentin, "and he pulled a shoe box from under his desk and there inside were a few of the toads! I was amazed to see a species that had been believed extinct." Quentin told me that Dr. Alcover was equally amazed. The two biologists stood enthralled, looking down at the elusive toads nestled there in the shoe box.

Quentin then met Dr. Joan Mayol and other Mallorcan scientists, who took him to see the place that they had earmarked as a site for a captive breeding program. "It did not seem to offer appropriate accommodations," Quentin told me, "and I suggested they might like to send a few individuals to the Jersey Wildlife Conservation Trust [now called the Durrell Wildlife Conservation Trust], which had a good record for breeding endangered species." Dr. Mayol readily agreed—but it was five years before the documentation was ready, since the appropriate authorities in Spain as well as Mallorca had to be approached. During this time, three other small populations of the toads had been found in the area.

Finally the conservation trust was able to send Simon Tong from the Herpetology Department to collect tadpoles for breeding in Jersey. They got their legs and lost their tails and all seemed to be going well—until they began to croak. "Every single one was a male!" said Quentin, laughing. The male has a ringing call to attract the females—apparently it sounds a little like a hammer hitting an anvil. For this reason, this toad is sometimes known as the *ferreret*, a Spanish word meaning "little blacksmith." And so those poor little ironworkers in Jersey croaked away in a futile attempt to summon nonexistent females! Fortunately the next consignment of toads soon arrived from Mallorca, this time with some adults—including females! After this, things went well and the toads prospered in their captive environment.

Since 1988, Quentin told me, several thousand have been successfully returned to Mallorca, both as adults and tadpoles, into areas known to be within the historical range of the species. Some 20 percent of the current population in the wild is derived from captive-bred stock that have been distributed in seventeen sites.

Of course, it is not all plain sailing. There are still threats posed by habitat loss and introduced species that prey on the toads and tadpoles (such as the viperine snake) or compete with them for food (like the green frog—which also eats them). More serious, perhaps, is a shrinking of water as a result of the numbers of tourists who visit the island. To address this there are plans to dam some of the toad's rivers to create suitable habitats. In fact, it was discovered by those working on the project that the toads love the granite water troughs made by the shepherds in the old days, placed in the deep shade so that they would not dry out.

In 2005, the dreaded chytrid fungus that has killed millions of amphibians worldwide was reported for the first time in Mallorca. It has so far only been found in two populations of the midwife toads. Fortunately, because they always live near torrents and only move up and down the stream where they are born, not from one stream to another, the virus was contained.

In 2002, it was decided that no more captive-bred toads or tadpoles should be sent back to Mallorca since there is little need and the potential cost—the risk of introducing disease—is huge. There is an educational program on the island, helping to raise awareness and instill pride in their unique, endemic toad. Already, as Quentin told me, "this toad has been the subject of a good many master's and a few PhD degrees."

The recovery program, supported by the Mallorcan government in collaboration with the Marineland Mallorca and Govern de les Illes Balears, is acclaimed as the model for amphibian recovery. It is the first amphibian species to have its original "critically endangered" status changed to "vulnerable." And when I visited Mallorca as part of a JGI-Spain lecture tour, I was able to congratulate government officials on this success. There is a new wave of concern for the environment and animal welfare in Spain as a whole, and this bodes well for the future of not only this endemic toad but other endangered wildlife as well.

Zino's Petrel *(Pterodroma madeira)*

This is a fascinating story in which a new species of petrel, believed to be extinct before it was even described, was rediscovered by Dr. Paul Alexander "Alec" Zino, a passionate amateur ornothologist. But for the determined efforts of Alec and his son Frank, Zino's petrels would indeed have slid into extinction.

These petrels are slender birds, with a body length of just over one foot and a three-foot wingspan. Like all petrels they spend months at sea, picking up food from the ocean surface with their short sturdy beaks. They breed on Madeira, a Portuguese island off the northern

The rediscovery and ongoing protection of Madeira's elusive petrel will forever be linked to the Zino family. Shown here is a historic photo of father Alec (left) and son Frank (right) working to find and protect the petrels on the Selvagem Islands in the 1980s. (Elizabeth Zino and René Pop)

Eventually, the third generation of Zinos picked up the work of monitoring and safeguarding the Zino's petrels. Shown here is grandson Alexander Zino, with chick. *(F. Zino of Freira Conservation Project)*

coast of Africa, arriving during the darkness of night, and flying up the steep valleys of the high mountains to their nesting sites among the sheer rock pinnacles. If there is no nest burrow available, the younger birds will dig new ones in which to lay their single eggs. About two and a half months after hatching, the fledglings launch themselves into the darkness; they will not return to Madeira for up to five years.

Our story begins in 1903, when a few dead birds were found and taken to Father Ernesto Schitz, a priest with a keen interest in natural history. He identified them—wrongly, it turns out—as Fea's petrels. Thirty years later those "Fea" skins were reexamined by petrel expert Gregory Matthews, who realized, to his excitement, that he was looking at the remains of a completely different species, one unknown to science. He named it *Pterodroma madeira*. Since there had been no reports of live birds since 1903, he assumed the species was extinct.

And then in 1940, a single dead petrel was found and taken, for identification, to Alec Zino. He immediately recognized that this bird was one of the new species described by Matthews: Clearly, it was not extinct after all! After this he and his son Frank made repeated trips to Madeira's high mountains where the birds were most likely to breed, listening for the calls of petrels. But they heard nothing and saw no signs.

Then Alec had an idea. Because this new species was so similar to the Fea's petrel in appearance, perhaps its call was similar too. He played recordings of Fea calls to shepherds in the high mountains—and one of them, Lucus, recognized the calls at once. He said they were "souls of shepherds who had died in the mountains." Lucus told Alec and Frank that they could hear those calls near Pico Cidrao, in the central massif.

And so in 1969, Alec, Frank, and Gunther "Jerry" Maul, a friend who had stimulated their fascination with petrels in the first place, drove to Pico Arcero, high in the mountains, then climbed down to a "stone table" where they huddled, waiting. Thinking back to that night Frank wrote: "It was bitterly cold and very dark; ideal for listening.

"Suddenly," Frank continued, "my father nudged me and said, 'Did you hear it?' We both listened all the more intently and heard this noise above that of the wind. 'Yes!' we both called in delight—waking Jerry, whose snoring we had been registering!!!" The "calls" stopped!! Soon, though (with Jerry wide awake from laughing), they heard the real calls and listened, entranced to the sounds that have been described (by ornithologist Malcolm Smith) as "ghostly nocturnal wailing."

Later that year a very small colony of the live birds was found, nesting on a rocky ledge. Apart from the local shepherds, Alec, Frank, and Jerry were probably the first people ever to see these petrels alive. For the next few years father and son returned during the breeding season to observe the birds. "It was not encouraging," Frank told me. "The breeding success at the known nesting burrows was terribly low."

During the season of 1986 they began systematic monitoring of the colony; at the one known nest ledge there were only six nests with eggs in them. And not one of the young birds survived the summer—almost certainly due to predation on eggs and chicks by rats. This finding was shocking, and it led to the launching of the first serious conservation organization, Freira Conservation Project (FCP), for predator control and systematic monitoring of the Zinos' colony.

"On September 12, 1987," Frank told me, "we pulled a ball of down out of a nest—the first chick we had ever handled!" They ringed it, returned it to its nest, and eventually it fledged. It was the only one that survived that year. However, as they persisted in their efforts to control the rats, things began to look brighter. And then, in 1992, just as they thought that they were winning the battle against rats, they lost ten birds to cats: "almost twenty-five percent of the known breeding population," said Frank.

In addition to baiting and killing rats, the new conservation group, FCP, then began trapping cats (since then about ten cats per year had been caught in the breeding grounds). As a result the breeding success of the petrels improved during following seasons. Nevertheless it would take years before numbers in the breeding colony increased, since each female lays only one egg, and each chick, after fledging, spends the next five years at sea.

A National Park and Hope for the Future

It was an exciting day when a team of FCP climbers discovered another small breeding colony. "The number of breeding pairs almost doubled overnight!" Frank told me. FCP then obtained funding to buy the breeding area from the private owners. And the government set aside a large area in the central mountains and laurel forests for a national park. Most important for the petrels, sheep and goats are no longer allowed to graze the high mountains. Fences were erected and shepherds whose flocks were excluded were compensated. This resulted in massive restoration of vegetation, much of which is endemic. It is believed that Zino's petrels used to nest in many other areas, and it is hoped that they will soon try new nesting sites. To encourage them, some artificial burrows have been constructed.

"Things are now running smoothly," said Frank, whose grown son Alexander and daughter Francesca are now involved in carrying on the family's protection of the Zino's petrel. In the 2008 breeding season there were about sixty to eighty nesting pairs. The Parque Natural de Madeira has taken on the conservation programs initiated by the FCP. And Frank wrote that "we even have eco-tourists coming to hear the birds at night." (How I should love to experience that myself!)

Frank ended by recalling "the huge honor that my father and I felt when the name Zino's petrel, suggested by W. R. P. (Bill) Bourne, stuck. It is very humbling and makes me all the more determined that all should go well for the future of this now less-rare species." One thing is certain: But for Alec and Frank, the Zino's petrel would be extinct, its eerie nocturnal calls silent forever.

The Large-Billed Reed Warbler *(Acrocephalus orinus)*

This little bird has been quietly getting on with its life not in a remote jungle but in the habitat around a wastewater-treatment plant outside

Bangkok! It was rediscovered in March 2006 by ornithologist Philip Round, who was making a survey there. Along with other, familiar birds, Philip captured a small warbler that he did not recognize. It had a long beak and short wings.

"Then it dawned on me—I was probably holding a large-billed reed warbler. I was dumbstruck," he said in an interview. "It felt as if I was holding a living dodo." The species had been identified and described in the Sutlej Valley of India in 1867; since then it had not been seen for 130 years. No wonder there was some debate as to whether this one specimen had been correctly identified. However, photographs and DNA samples subsequently confirmed the identification. The large-billed reed warbler is one more species that has defied extinction.

This rediscovery, of course, was very exciting to ornithologists and the bird was a hot topic of conversation in their circles. Probably this is why, just six months later and while biologists were still investigating the waste-water plant birds, another specimen was found. This one was dead—discovered in the UK in a drawer in the Natural History Museum at Tring. There, for more than a hundred years, it had been lying with other reed warblers collected from Uttar Pradesh in India in the nineteenth century. It, too, was confirmed as a large-billed reed warbler through DNA analysis. Ornithologists are now speculating that other populations of the bird may yet be found in Thailand, and perhaps also in Burma or Bangladesh.

The Caspian Horse

This story is about a very small and very beautiful breed of horse, and an American woman, Louise, who "discovered" and rescued them from obscurity in Iran. Louise married a young man from the Iranian royal family, Narcy Firouz, and became a princess. In 1957, the young couple established the Norouzabad Equestrian Center, where the wealthier Iranian families sent their children to learn to ride. The trouble was that all the typical horses of Iran—the Arabian and Turkoman—were too big for the smaller children, including their own three. And so when, in 1965, Louise heard rumors of a small pony in the Elburz Mountains near the Caspian Sea, she decided to investigate. She set out on horseback with a few women friends—it was not usual for women to travel like this, and the journey (the first of several she would make) was potentially dangerous. But all went well, and she found the "ponies." They were being used as work animals, pulling carts, malnourished and covered with ticks.

Almost at once Louise realized that these were not ponies at all—

Louise Firouz "discovered" and rescued the Caspian horse from obscurity in Iran. Shown here is Fereshteh, the first foal born after the Islamic Revolution. Tragically, during the revolution most of the Caspian horses were lost—auctioned as beasts of burden or slaughtered for meat. *(Brenda Dalton)*

they had the distinctive gait, temperament, and unique facial bone structure of horses. Very small, narrow horses to be sure, standing only about 11.2 hands high (one hand is four inches), but horses for all that.

As she pondered the nature of this little horse, Louise suddenly remembered seeing, on the walls of the ancient palace in Persepolis, rock relief carvings of a horse that looked very much like the one she had just found. The Lydian horse depicted in those carvings had the same small, prominent skull formation. With a sense of excitement, Louise began to wonder whether, hidden beneath the matted coats of the work animals she had found, was a true representative of the ancient lost breed of the royals. The more she thought about this, the more certain she became.

The Lydian horse had been used for chariot racing and in battle, a suitable gift for kings and emperors. It was thought by many to have been the ancestor of the Arabian—and it had been thought extinct for a thousand years! Louise found that there were still five purebred horses in the village, and she bought three of them. After extensive DNA testing, archaeozoologists and genetic specialists agreed with Louise that these little horses were indeed the ancestral form of the Arabian. What an incredible find!

Louise made other excursions to the region, trying to find out how many of the little horses remained. I spoke with Joan Talpin, a close friend of Louise, who went with her on several of those searches. She told me the villagers were always friendly, and she remembers how the owners of the tiny inns where they stayed would go out to cut straw for fresh sleeping mats so that the visitors would not be plagued by bedbugs or fleas! In the end, Louise estimated there were about fifty of the horses, which she called Caspians, along the southern coast of the Caspian Sea. She purchased a few more, Joan told me—six stallions and seven mares—to found a breeding herd. Louise's favorite remained that very first horse she found, whom she named Ostad Farsi for the professor. "He was a true gentleman," said Joan, "and the breed owes much to him." He was also loved by Louise's children, who spent hours riding him and the other rescued Caspians.

At first Louise and her husband, Narcy, financed the breeding themselves, but then in 1970 a Royal Horse Society (RHS) was formed in Iran. The society's mission was to protect and maintain Iran's native breeds, and it bought all Louise's Caspians, by then numbering twenty-three. Louise and Narcy then started a second private herd near the Turkmenistan border. When two mares and a foal were killed by wolves, Louise, wanting to ensure that some of the horses be kept safe, arranged for eight of them to be exported to Britain in 1977. The RHS was angered—presumably they had not been consulted—immediately banned all further exports of Caspian horses, and began collecting up all of the little horses that remained in Iran, including all but one of the Firouzes' second herd.

Surviving Revolution and War

Then came the 1979 Islamic Revolution. The Firouzes, because of their connections with the royal family, were arrested and imprisoned. Narcy was jailed for six months but Louise only for a few weeks, for she remembered advice given to her by a friend—that if she went to prison, she should go on a hunger strike. This worked—but, Joan told me, "Louise was thin anyway and must have been a beanpole when she came out!" Tragically, during that time most of the Caspian horses were lost—auctioned for use as beasts of burden or slaughtered for meat.

Louise, however, was a survivor—and she was passionate about saving and protecting the bloodline of her beloved Caspian horses. She managed to rescue some of those that remained from starvation and slaughter and established, for the third and last time, a small herd—to try to save the breed from extinction in Iran. And once again, before this was banned by the new government, she managed to export some of them to safety. The last such effort was in the early 1990s, when she sent seven horses on a tortuous and dangerous journey to the UK. They had to pass through the Belarus war zone, where bandits attacked and robbed the convoy. The horses arrived safely, but it had been a costly business. Soon after, in 1994, her husband died, and Louise could no longer afford to continue with her breeding program in Iran. She sold the remainder of her herd to the Ministry of Jehad, but was often called upon to advise on their management. She also assisted John Schneider-Merck, a German businessman, to establish his own small private herd of Caspians in Iran.

The Future of the Caspian Horse Ensured

With Iran's many political upheavals—the overthrow of the shah during the Islamic Revolution, bombing during the Iran–Iraq War, the very real threat of famine—as well as the Caspian's former association with royalty, the fate of these horses was ever in the balance. One moment they were considered a national treasure, the next they were seized as wartime food. But thanks to Louise, who had exported a total of nine stallions and seventeen mares, the future of this ancient line has been ensured. Today they can be found in England, France, Australia, Scandinavia, New Zealand, and now the United States.

Much of the history of this little horse can be found in *The Caspian Horse*, written by one of Louise's close friends, Brenda Dalton. She writes that Caspians are "one of the oldest and most gentle breeds in the world. They become attached to you, and are more dependent on us, more 'doglike' than other breeds of horses or ponies. They are very charismatic and very, very pretty and very engaging." But for Louise, they would almost certainly have vanished without a trace. The fact that she "discovered" them, before it was too late, must have given her great joy. Later she would say that after finding the first Caspians, she watched the ancient horse "trot serenely back into history."

Louise, "Iran's lady of horses," died in May 2007, and when I spoke to Brenda on the phone she had just returned from a memorial service held in the UK. What a fascinating and amazing person, what an extraordinary life. Above all she understood and loved horses, and she must have suffered greatly when her beloved Caspians were sold back into drudgery and for slaughter. But despite the setbacks, and as a result of her courage and determination, she saved a rare and charismatic breed, reintroduced it to the horse-loving world, and became, herself, an integral part of its history.

The East London Fish: The most startling "living fossil" ever discovered.

THE OUTDOOR WORLD

A LIVING FOSSIL CAUGHT IN THE SEA

Amazingly, we are even rediscovering species from the distant prehistoric world—once believed to only exist as fossils. Shown here a news clipping about the coelacanth, an animal species that has survived, unchanged, for sixty-five million years. *(South African Institute for Aquatic Biodiversity)*

Living Fossils:
Ancient Species Recently Discovered

Imagine finding a living species previously known only from fossils! A species from an ancient prehistoric world that has existed, beyond our knowledge, for millions of years. The coelacanth, an enormous shark-like fish, was discovered just before World War II. Because I was only four years old, it was not exciting to me at the time. It is very exciting to me now. An animal species that has survived, unchanged, for sixty-five million years! And no one knew about it—except, I suppose, fishermen who had occasionally caught one in their nets, and they would have had no idea that it was anything untoward. It was indeed known to science, but in the form of fossils, stored in various museums, of little interest to any save those paleontologists who happened to be interested in fish. For them, the discovery was as though a living dinosaur had been found!

When I worked with Louis and Mary Leakey at Olduvai in 1958, I would sometimes stand, holding the fossilized bone of some long-gone species, and imagine how it would have looked in life. Indeed, it sometimes led to near-mystical experiences. As when I found the tusk of an extinct giant pig and seemed suddenly to see it standing there, huge and fierce. Saw its coarse brown hair, the crest of black hair along its back, its bright fierce eyes. I seemed to smell the animal, hear it snort. And then it was gone and I was left looking down at a piece of prehistoric ivory, slowly returning to reality.

The coelacanth comes from a far more ancient era than that pig. It is as though one of the fish, from those prehistoric seas I had longed to visit as a child, has come swimming into the present. And I can so easily imagine the overwhelming feeling of excitement of the scientists who handled and studied that first coelacanth. Indeed, they must sometimes have imagined they were dreaming.

The Wollemi pine was also known only from the fossil record—from imprints of its leaves on ancient rock. And it, too, dates back sixty million years. When the first specimen was picked from a tall tree in a remote and unexplored canyon in Australia, the biologist who found it had no idea that he had made a major discovery, that he would have the extraordinary honor of having a "living fossil" named for him. Indeed, it took a long time and many hours of discussion and searching through herbarium specimens before its true identity was finally revealed. That was truly the botanical discovery of the last century, just as the coelacanth was one of the major discoveries in the animal kingdom. The future of the tree is assured—that of the fish is uncertain. The stories of both are fascinating.

The Most Beautiful Fish or "Old Fourlegs"
(Latimeria chalumnae)

Toward the end of 1938, Marjory Courtenay-Latimer, a twenty-three-year-old museum curator in East London, South Africa, noticed a very strange-looking fish in the catch of the trawler *Nerine*. She often went to look at the sea life brought in by the fishermen, but she had never seen anything like this before. In an interview, she said it was "the most beautiful fish I have ever seen, five feet long and a pale mauve blue with iridescent silver markings." She and the museum staff knew that it was unique and of great scientific value. She preserved as much of the fish as possible, drew it, and sent the now famous sketch to renowned ichthyologist Professor J. L. B. Smith.

I would love to have been there when, finally, Professor Smith and the remains of that fish got together. Already there was speculation as to the identity of the deep-sea creature—and early in 1939, Smith announced to a stunned world that it was a coelacanth, a fish previously known only from the fossil record. It had been considered extinct for some sixty-five million years.

For the next fourteen years, no more coelacanths were reported, but then, in 1952, one was found in the Comoros. Professor Smith—I imagine with much excitement—went to fetch it. This find was considered so important that the then prime minister, Dr. D. F. Malan, allowed

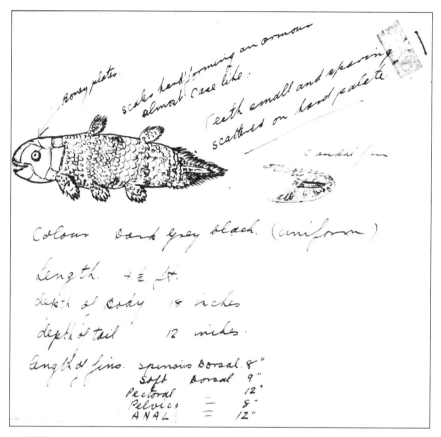

In 1938 Marjory Courtenay-Latimer, a twenty-three-year-old museum curator in East London, South Africa, saw a strange-looking fish in the catch of a local trawler. She drew the fish, and sent this now famous sketch to renowned ichthyologist Professor J. L. B. Smith, who identified it as a coelacanth, a sixty-five-million-year-old species. *(South African Institute for Aquatic Biodiversity)*

him to use a Dakota of the South African Air Force to transport the fish back to East London! More scientists became interested, and more attempts were made to try to see these fish in their natural habitat. And then came the first amazing footage of coelacanths swimming in the ocean. It was shot from the manned submersibles *Geo* and *Jago* by Professor Hans Fricke and his team.

Coelacanths are large fish growing to about six feet in length; the heaviest recorded so far was 243 pounds. Professor Smith wrote a book about them, which he titled *Old Fourlegs*—a reference to the lobed fins

Historical photo of Marjory Courtenay-Latimer with a mounted coelacanth. *(South African Institute for Aquatic Biodiversity)*

that he and other scientists thought might be precursors to the arms and legs of land vertebrates.

Recently I was in touch with Dr. Tony Ribbink in Grahamstown, South Africa. He is the CEO of the Sustainable Seas Trust, founded to study and protect endangered species in the ocean canyons and caves of Kenya, Tanzania, Mozambique, Madagascar, the Comoros, and South Africa. He got involved with coelacanth research and conservation in 2000 when scuba divers discovered a colony in the Saint Lucia Wetland Park off Sodwana Bay, South Africa. They were more than a hundred yards deep when they found and filmed coelacanths in canyons about two miles from the shore.

"The discovery of the coelacanths in a marine park and world heritage site," he said, "was a wake-up call." He likened it to finding elephants in a terrestrial park years and years after the park had been established. I asked if he had seen coelacanths in the wild. "Yes I have," he told me, "at depths from 105 to over 200 meters. They are amazing—very quiescent, very tolerant of each other, slow moving and mystical."

The South African Institute for Aquatic Biodiversity launched the African Coelacanth Ecosystem Programme, which works in Comoros,

Kenya, Madagascar, Mozambique, South Africa, and Tanzania. They have engaged hundreds of researchers, students, and public officials from nine countries and gradually gained new insights into the ecology, distribution, and behavior of these amazing survivors from ancient times. But still many of the fundamental questions, asked initially in the late 1930s by Marjory Courtenay-Latimer and Professor Smith regarding life history, breeding behavior, gestation period, where the young are born, whether parental care is practiced or whether the young hide until they are large enough to join adult groups, remain unanswered. No one has ever knowingly seen a young coelacanth in the wild.

"When our research began in 2002," said Tony, "only one coelacanth was known from Mozambique, one from Kenya, four from Madagascar, some from Comoros, and we know that our South Africa population has at least twenty-six individuals."

In 1979, a coelacanth was found off Sulawesi by an Indonesian fisherman. This turned out to be a different species, *Latimeria menadoensis*. Another of these, again off Sulawesi, was caught alive in 2007 and actually lived, in a quarantined pool, for seventeen hours.

Tragically, these living fossils—which have survived innumerable stresses over the millennia yet remained essentially unchanged—are now vulnerable to extinction. This is because while they are fairly unpalatable and are not targeted by fishermen, they are caught accidentally as a bycatch. Increasing demand for fish and a depletion of the inshore resources have seen fishermen move into deeper water to set gill nets, thus penetrating the habitats of the coelacanth around Africa and Madagascar. The first coelacanth bycatch recorded in Tanzania was in September 2003; since then, nearly fifty have been caught. All have died. This represents the greatest known rate of coelacanth destruction anywhere.

Fortunately the Tanzanian authorities, with the help of the Sustainable Seas Trust, are planning to develop, off the coast of Tanga, one of several marine protected areas. These refuges are not exclusively for coelacanths but are part of a plan to protect special offshore ecosystems while working out sustainable ways of harvesting them to benefit coastal human communities as well as the fish. But the coelacanth is of such importance that a major awareness campaign has been launched to let the people know about the extraordinary prehistoric fish in their waters.

"Coelacanths are rare, beautiful, and intriguing," says Tony. "They have brought together people of many cultures and countries, and inspired a more harmonious relationship between us and the rest of the

A fossilized branch alongside a recent clipping of the rediscovered Wollemi pine that belongs to the two-hundred-million-year-old Araucariaceae family. (© J. Plaza RBG Sydney)

living world. To the countries of the Western Indian Ocean, they are an icon for conservation—the panda of the sea. And a symbol of hope."

A Noble Discovery: The Wollemi Pine *(Wollemia nobilis)*

On Saturday, September 10, 1994, David Noble, a New South Wales national parks and wildlife officer, was leading a small group in the Blue Mountains of Australia about a hundred miles northwest of Sydney, searching for new canyons. David has been exploring the canyons of these wild and beautiful mountains for the past twenty years.

On this September Saturday, David and his party came across a wild and gloomy canyon that he had never seen before. It was hundreds of yards deep, the rim fringed by steep cliffs. The party abseiled down into the abyss, past numerous small waterfalls of sparkling water. They swam through the icy waters, and then hiked through the trackless forest. During this adventure, David noticed a tall tree with unusual-looking leaves and bark. He picked some of the leaves and put them in his backpack, then forgot about them until he got home and retrieved a slightly crushed specimen. He first tried to identify it himself but could not find anything to match. He had absolutely no idea that he had just made a discovery that would astound botanists and enthrall people all over the world.

Unraveling a Mystery

When he showed the battered leaves to botanist Wyn Jones, Wyn asked if they had been taken from a fern or a shrub. "Neither," David replied. "They came from a huge, very tall tree." The botanist was puzzled. David helped in the search that followed, looking through books and on the Internet. And gradually the excitement grew. As the weeks went by, and the leaves could not be identified by any of the experts, enthusiasm grew even more.

Eventually, after many botanists had pored over David's leaves, it became clear that the tree was a survivor from millions of years ago—the leaves matched spectacular rock imprints of prehistoric leaves that belonged to the two-hundred-million-year-old Araucariaceae family.

Clearly it was necessary to find out a good deal more about this extraordinary tree, and David led a small team of experts back to the place where the momentous discovery had been made. As a result of that expedition, and exhaustive research into the literature and examination of museum samples, the tree, a new genus, was named, in honor of the finder, *Wollemia nobilis*, the Wollemi pine. It struck me, as I was talking to David, that for the sake of the majestic tree it was lucky that David had an appropriately majestic name. After all, it could have been found by a Mr. Bottomley!

It is indeed a noble tree, a majestic conifer that grows to a height of up to 130 feet in the wild, with a trunk diameter of more than 3 feet. It has unusual pendulous foliage, with apple-green new tips in spring and early summer, in vivid contrast with the older dark green foliage.

Continuing research showed that the pollen of this new tree matched that found in deposits, across the planet, dating from the Cretaceous period somewhere between 65 and 150 million years ago when Australia was still attached to the southern super-continent of Gondwana. One professor of botany, Carrick Chambers, the director of the Botanic Gardens Trust–Sydney, exclaimed in wonder: "This is the equivalent of finding a small dinosaur alive on earth."

All this for a pine tree. Horticulturalist dangles from a helicopter to collect the seeds of the prehistoric Wollemi pine. (© J. Plaza RBG Sydney)

Their Secret Home

It is now known that there are a few small stands of these rain forest giants in that canyon, all part of the same population of less than a hundred individuals. Very few people—a handful of scientists only—have actually been to see the trees growing in the wild. The exact location has been kept a strict secret to try to protect these ancient trees from new diseases. This is very important, as there is an unprecedented total lack of genetic diversity among Wollemi pine individuals. On one of botanists' more recent visits, it was found that a ground fungus that attacks the roots of trees had invaded the canyon, perhaps taken there by a bird or by the wind. Immediately measures were

David Noble shown here with a Wollemi pine at Mount Annan Botanical Gardens in Sydney, Australia. To this day David keeps the exact location of his original discovery a secret—telling only a few scientists and horticulturists. (© *Botanic Gardens Trust, Simone Cottrell*)

taken to treat the ground in the vicinity of the precious Wollemi pines to eliminate the danger.

Investigations of the rings of the trunk show that the Wollemi pine has withstood a range of potentially lethal environmental conditions, including forest fires and windstorms, and lived through extremes of temperature—from 104 degrees Fahrenheit on the one hand to 10 degrees Fahrenheit on the other. In freezing weather, the growing tips are sealed with caps of resin, which is probably what enabled the Wollemi to survive no fewer than *seventeen* ice ages! The trunks have unusual bubbly bark—"a bit like Coco Pops," said Dave.

"Every Leaf Precious—A Seedling Priceless"

It was obviously important to try to propagate the Wollemi pine to ensure its survival into the future lest anything should happen to the wild individuals. In an *Australian Geographic* that appeared in 2005, John Benson, senior ecologist at the Botanic Gardens Trust–Sydney, is quoted as saying: "We caught a species at the point of evolutionary death. But the species isn't going to go extinct, no. We have come in and played God."

A huge effort is under way to propagate and commercialize these pines, not simply as a safeguard for the species itself but also to raise money for conservation of these and other endangered plants. The work began in 2000 and goes on, behind closed doors, in the Wollemi nursery compound at Gympie. This is where Lyn Bradley has been working since the start of the program.

"Initially," she said, "every leaf was precious, a seedling was priceless. Now there are hundreds." She is absolutely committed to this work, and passionate about the pines—to some of which she has assigned pet names. She and her boss, Malcolm Baxter, are the only two who know the secrets of the pine's commercial propagation, and both feel privileged to be involved with this extraordinary species. They hope, among other things, that by propagating the pines and selling them to botanists, gardeners, and collectors around the country, people will be less desperate to visit the canyon to see the trees in the wild—but I doubt it. I saw one of the two that was donated to Kew Botanical Gardens during my recent visit there. It was planted by Sir David Attenborough and is growing splendidly within its protective iron cage. And in Australia, I had the privilege of planting one of the little saplings on the grounds of Adelaide Zoo.

I am, of course, delighted to have seen and even handled living tissue descended from the ancient giants. But it does not stop me longing to visit that dark and mysterious canyon that has, for millions of years, hidden its secret, and stand in the presence of the original trees themselves. Indeed, several of those privileged few who visited the canyon in the early days said that the experience was close to spiritual. Long may they stand there, undisturbed by the frenzy of a modern world, so different from anything they have known and endured for so many millions of years.

PART 6

The Nature of Hope

It was a joyous moment, releasing this trout that was quivering with life into the cold, cleansed water of a once-polluted stream. Sudbury, Ontario, Canada. (David Wiewel)

Healing Earth's Scars:
It's Never Too Late

Throughout the pages of this book, we have shared stories of species that, although rescued from the brink of extinction, are still endangered by lack of suitable habitat in the wild. Tropical and old-growth forests, woodlands and wetlands, prairies and grasslands, moorlands and deserts—all landscapes—are disappearing at a terrifying rate.

So how, people ask, can I have hope for the future? Indeed, I am often accused of being unrealistically optimistic. What is the point of saving endangered life-forms, people ask, if there is nowhere for them to live except in zoos? So let me share why it is that, against all odds, I have hope for the animals and their world. Why it is I believe that human know-how and the resilience of nature, combined with the energy and commitment of dedicated individuals, can restore damaged environments so that, once again, they can become home to many of our endangered species.

My four reasons for hope, about which I have written and spoken extensively, are simple—naive perhaps, but they work for me: our quite extraordinary intellect, the resilience of nature, the energy and commitment of informed young people who are empowered to act, and the indomitable human spirit. When human know-how and the resilience of nature are combined with the resourcefulness of dedicated individuals, desecrated landscapes can be given another chance—just as animal and plant species can be saved from extinction.

We have already discussed the restoration of island habitats. Now let me share some of the successful projects that have restored mainland habitats, including streams, rivers, and lakes. Some of these efforts were undertaken with the express intent to save endangered wildlife. In some cases, cleanup efforts were initiated by the government, in others by citizens determined to create a better environment for themselves and their children. A businessman whose operation had caused horrible ecological damage suddenly felt he must put things right; a child made a pledge to restore a mountain—and made his dream come true. All of these efforts are described more fully and illustrated on our Web site (janegoodallhopeforanimals.com).

Kenya Coast: From Wasteland to Paradise

One quite extraordinary project resulted in the transformation of a five-hundred-acre "wasteland," created by twenty years of quarrying by the Bamburi Portland Cement Company, into lush forest and grassland. And the project was initiated—in 1971—not by a group of concerned environmentalists but by Dr. Felix Mandl, the man whose company had caused the devastation. The miraculous change was brought about by the company's remarkable horticulturist, Rene Haller.

When Rene began, the site appeared as "a monstrous lunar-like scar on the landscape, barren, desolate and exposed to the hot tropical sun." The task seemed all but impossible. "It was appalling to note that even in the oldest parts of the quarry no plants had been able to establish themselves," wrote Rene. "I spent countless agonizing hours in the hot and dusty barren land, found a few ferns and perhaps half a dozen tiny bushes and grasses which were struggling to take root, sheltering behind some of the remaining rocks. It was hardly an encouraging environment for tree planting."

Yet today the area is a self-sustaining habitat for wildlife, including thirty species of animals and plants that are on the IUCN endangered species list. And in addition to recreational facilities for visitors, there are countless environmentally sustainable opportunities to improve the lives of the local people. It has become a major environmental education center for Kenya, and is used by schools throughout the country.

From the very beginning of the project, Rene had held the firm belief that, if he looked hard enough, nature would provide the solutions to all his problems. The description of how he tackled his enormous task, step by step, learning from nature, introducing each new species with care, is incredibly interesting and inspirational. It is living proof that the rehabilitation of a man-made wasteland is not only possible, but can be accomplished with sound organic principles.

The Man Who Restored Forests to a Mountain

This story—one of my favorites on our Web site—is about the absurd dream of a six-year-old boy that eventually came true. There was no fairy godmother waving a magic wand—only his sheer determination to make his childish vision into reality.

This hero is Paul Rokich. His father worked for the big copper mine at the foothills of the Oquirrh Mountains in Utah. Paul remembers standing with his father in 1938, when he was six years old, and looking up at the mountains. They were black, the once beautiful forests (that he had seen in a photograph in a school textbook) gone, destroyed by logging, by extensive sheep grazing, and finally by the toxic emissions of the smelting operations.

Paul told his father that one day he would go up those mountains and put the trees back. Surely an impossible task. Yet twenty years later, he set to work to honor his pledge. Every evening, every weekend, year after year, he carried buckets of grass seed up the mountain, driving as far as he could then walking—and sowing. For fifteen years, Paul worked mostly alone, with his own money. Sometimes his family and friends helped. And despite the countless setbacks and disappointments he endured, he never gave up.

Finally the Kennecott Company was shamed into cleaning up the poisonous emissions from its smelting operations, spending millions of dollars. And eventually company managers hired Paul to help them with their belated restoration project. Today the Oquirrh Mountains are green, covered with native grasses and plants originally seeded by Paul, and trees that he planted as seedlings. And the animals have returned.

I have flown over those mountains, looked down on those trees—and marveled. Paul sent me a laminated leaf from one of the very first trees he planted. I carry it around the world, for it symbolizes both the indomitable human spirit and the resilience of nature if given a helping hand.

Sudbury, Ontario

When I first visited Sudbury in the mid-1990s, I heard an extraordinary story that illustrates how a vast landscape utterly devastated by years of destructive human activity can—with time, money, and determination—recover. It is one of the largest community-based environmental restorations of industrially despoiled land ever undertaken. The full story, on our Web site, is amazingly inspirational and one that I never tire of sharing.

It tells how irresponsible logging and industrial pollution gradually created a landscape similar to the surface of the moon, and how the citizens eventually determined to do something about it. I found it so inspiring that I returned, several years later, to learn more. I walked through a glorious landscape where young trees were bursting into spring glory, flowers bloomed everywhere, and the air was full of birdsong. It was almost impossible to believe that, not so long ago, everything had been barren and lifeless—but one area has been left untreated, and the blackened rock is a stark reminder of the harm our species is able to inflict.

The original forests have not returned, nor will they. But the area is beautiful, and much of the wildlife is back. As I turned away from the blackened rocks of yesterday, I was just in time to glimpse the arrow-swift flight of a peregrine falcon—back again after more than fifty years. It was almost as though nature herself was sending me a message of hope to share with the world. They gave me a feather, found near one of the three peregrine nests, as a symbol of all that can be done to heal the scars we have inflicted on Planet Earth.

Before I left Sudbury, I had the joy of releasing a brook trout into the clean water of a stream that had, until recently, been dank, poisoned, lifeless.

Water Is Life

The pollution of our streams, rivers, lakes, and oceans is one of the more shocking results of the use of chemicals and other damaging agents in agriculture, industry, household products, golf courses, and gardens, since much of this poison is washed into the water. Even many of the great aquifers are now polluted. This chemical pollution has led to the destruction of many endangered species' habitats. Yet there are signs of hope here: Slowly our waterways are being cleaned.

I remember when the River Thames in London seemed beyond hope, flowing through London lifeless, contaminated, and murky. Fifty years ago, the Potomac River passed through Washington, DC, stinking like a sewer. And many other major waterways were in much the same state as so many of those in China today. In the United States, Lake Erie was at one time declared a fire hazard, and the Cuyahoga River actually went up in flames and blazed for at least two days! Of course, most species of flora and fauna vanished from such contaminated waters.

Today, however, many of these rivers and lakes have been cleaned up—often at huge expense—and much of the wildlife has returned. A couple of years ago, for example, bass fishing opened up in the Potomac, a clear indication of much cleaner water. Fish are thriving in at least parts of Lake Erie. And fish are back in the River Thames, where water-birds are once more breeding.

Here I want to mention just a few of the water cleanup projects that have come to my attention, many of which were undertaken in order to protect fish on the endangered species list.

How a Fish Led to the Cleaning of the Hudson River

Thirty years ago, the Hudson River and its surrounding waterways were so polluted that its population of short-nosed sturgeon became the first fish species to be listed (in 1972) as endangered. This resulted in a massive effort to clean up the river. Over the past fifteen years, the population of these fish in the Hudson River (next to one of the busiest cities in the world) has increased by more than 400 percent. The Manhattan area has the most urban estuaries on the planet, so the cleaning of its waters is a major conservation success story. Indeed, the environment has been so improved that there are even plans to introduce oyster reefs and shoreline wetlands in Harlem!

The Amazing Return of the Coho Salmon

In the 1940s, coho were so abundant in California rivers that their numbers were estimated at two to five hundred thousand statewide. And as recently as the 1970s, California's coho fishery still pulled in more than seventy million dollars a year in revenue. But since 1994, commercial fishing for coho has been completely shut down and the fish is listed both state and federally as endangered. It was because of this dramatic decline that a coalition of conservation partners—including landowners and industry—began working to monitor and restore the health of the Garcia watershed, clogged by sediments resulting from irresponsible logging practices.

I happened to be in town when the *San Francisco Chronicle* published an article giving good news. While snorkeling in the headwaters of the

Garcia River, Jennifer Carah, a scientist with the Nature Conservancy, and Jonathan Warmerdam from the North Coast Regional Water Quality Control Board, spotted juvenile coho salmon.

I called Jennifer, and she told me that since then young coho had been spotted in five of the twelve sub-watersheds in the river basin. In many of these streams, they had not been seen since the late 1990s. It was an exciting time—Jennifer told me that when she identified those young coho, she "squealed so loudly that Jonathan heard the sound even though we were both underwater"!

There are other great stories, like the demolishing of a lakeside resort to save a minnow-size fish in Nevada, and building an area of wetlands so that carefully selected plants could clean the polluted water of a river in Taiwan. These and other accounts are detailed on our Web site (jane goodallhopeforanimals.com).

Fortunately the looming threat of global water shortage has been acknowledged, and many of the stories in this book describe the efforts of those who are fighting against the reckless use of water for agriculture, industry, and domestic applications, the pollution of rivers and lakes, the draining of the wetlands, and so on.

Today we fight wars about oil, but as Ismail Seregeldin (then with the World Bank) said at the end of the last century: "The wars of the next century will be fought over water." We *could*, with major changes in the way most people live today, survive without oil. But we *could not* survive without water.

Hope for China

Almost always, when I voice my hope that we humans can find a way out of the environmental mess we have made, someone will point out what is happening in China. Do I realize, they want to know, the extent to which that giant country, containing one-fifth of the world's human population, is destroying its environment? And the threat that this poses for the rest of the world? I do, indeed. I have been to China once a year since 1998 and seen with my own eyes the speed of development, the staggering number of new roads and buildings—and cities—that spring up almost overnight. And I know full well that this rapid economic development has taken a heavy toll on the environment. In many cases, it has led to a great deal of human misery also.

As China opened up in the early 1980s, people were offered jobs manufacturing goods for outside markets—and the biggest migration in history was set in motion as the rural poor flocked to the new cities. And there, only too often, they found themselves and their children working in sweatshops, exploited so that China could undercut prices of goods made in the West. They tolerated this because they believed or hoped that it would eventually create a new economy from which they could benefit.

Meanwhile, the level of environmental degradation has soared. Two-thirds of China's main rivers are too polluted for the water to be used for drinking or agriculture. The aquatic ecosystems have been destroyed—the Yangtze River dolphin became extinct. There has been devastating destruction of habitats across the country. And having harmed so much of her own environment, China, desperate to acquire materials such as timber and minerals to sustain her economic growth, is plundering the natural resources of other countries. Especially in Africa where many politicians are willing to sell off the future of their children to make a quick buck.

No wonder so many have given up on China's environment—including many of the Chinese people. But it is important to realize that China is only doing what has been and often still is being done by many other countries. The impact is worse because of the country's staggering number of people and, until fairly recently, the government's refusal to admit there was anything wrong.

The good news is that people in China are now beginning to talk openly about the need to improve the environment and to set aside areas for the conservation of wildlife. (See this book's chapters on the giant panda, the crested ibis, and the milu.) Another story, highlighted on our Web site, describes steps being taken to preserve areas of wetland to benefit the critically endangered Chinese alligator. Moreover, JGI's youth program, Roots & Shoots, which involves young people of all ages in activities to improve the environment for wildlife as well as their own communities, is active in many parts of the country, with offices in Beijing, Shanghai, Chengdu, and Nanchang. There are about six hundred groups in all.

And the story of the Loess Plateau is another reason for hope. It is an area approximately the size of France in the northwest of China. It is home to about ninety million people who were, for many long years, trapped in a vicious cycle of poverty and environmental destruction that only got worse as time went on. For years, the Loess Plateau was considered the most eroded place on earth.

The almost miraculous restoration of this desolate area to a landscape boasting a thriving environment for people and at least some animals has been documented by my friend John Liu in his inspirational film *Earth's Hope*. It illustrates what can be done when a powerful government, backed by the World Bank, decides to take action.

Clearly the hundreds of millions of dollars spent were a wise investment, for already the local communities are thriving. The sense of hopelessness once shared by the population has been replaced by cautious optimism, and young people now expect an education and a future.

And there is hope for wildlife, too. It was decided from the start that there should be clear distinction between land designated for human use, and land that would be most valuable set aside to ensure, for example, protection of the watershed, soil stability, carbon sequestration, and biodiversity. And this "ecological land" could provide a refuge for local endangered species—rescuing them from the extinction many are currently facing.

Lesson from Gombe

The extreme environmental degradation of the Loess Plateau came about because the people sank ever deeper into poverty and hopelessness. Again and again I have seen, as I travel around the developing world, how rural poverty (that so often goes hand in hand with overpopulation) almost invariably causes great damage to the environment. But it was in Tanzania that I suddenly realized that we could only save the Gombe chimpanzees and their forests, in the long term, with the support of the local people. And that we could not hope for such support while they themselves, desperately poor, were struggling to survive.

When, in 1960, I arrived at Gombe National Park to start my chimpanzee study, lush forest stretched for miles along the eastern shores of Lake Tanganyika and inland as far as the eye could see. But over the years, growing populations of local people, swollen by refugees, cut down the trees for firewood and building poles. By the early 1990s, the trees outside the park had almost all gone, and much of the soil was exhausted. Women had to walk farther and farther from their villages in search of fuel wood, adding hours of labor to their already difficult days.

Looking for new land to clear for their crops, people turned to ever steeper and more unsuitable hillsides. With the trees gone, more and more soil was washed away during the rainy season; the soil erosion worsened and landslides became frequent.

By the late 1970s, the chimpanzees were more or less trapped within their tiny thirty-square-mile national park. There could be no exchange of females between groups—which prevents inbreeding—and with only some one hundred individuals remaining, the long-term viability of the Gombe population was grim. Yet how could we even try to protect them while the people outside the park were so desperate, envious of the lush forested area from which they were excluded?

Building Up Goodwill

Clearly it was necessary to gain the goodwill and cooperation of the villagers. In 1994 the Jane Goodall Institute (JGI) initiated TACARE (take care), a program designed to improve the lives of the people in these very poor communities. Project manager George Strunden put together a team of talented and dedicated local Tanzanians who visited the twelve villages closest to Gombe to discuss their problems. They worked out together how TACARE could best help.

Not surprisingly, conservation issues were not listed as top priorities. The main concerns were health, access to clean water, growing more food, and education for their children. And so, working with regional medical authorities, we introduced a new level of primary health care in the villages, including basic information about hygiene and HIV-AIDS. We established tree nurseries and developed ways to restore vitality to exhausted land—farming techniques best suited for the degraded soil. Roots & Shoots, our educational program for youth, was eventually introduced into all the villages. And as TACARE became ever more successful, we were able to start a micro-finance program enabling women to take out very small loans (almost always repaid) to start their own projects—which have to be environmentally friendly and sustainable.

The Importance of Women

All around the world, it has been shown that as women's education improves, family size tends to drop—and after all, it was the growth of the population in the area that first led to the grim conditions TACARE was trying to address. It would be irresponsible to introduce ways of growing more food and saving the lives of more babies, without, at the same time, talking about the need for small families. There are TACARE-trained volunteers from each village, men as well as women, who provide counseling—that is well received—about family planning.

Information about family planning, along with access to health care for her children, enables a woman to realistically plan her family. If she has also received an education, things will go even better. So we started a scholarship program for girls—for a poor family is more likely to educate boys, leaving the girls, once they have finished their first years of compulsory education in the primary schools, to help at home. Some of our girls are now in college.

Restoring and Protecting

Recently I went with our forester, Aristedes Kashula, to one of the villages. A woman demonstrated her new cooking stove, which greatly reduces the amount of firewood she needs. Because all the women get fuel wood from a village woodlot of fast-growing trees, they no longer need to hack at the stumps of trees that once grew on the bare hillside. And such is the regenerative power of nature that a new tree will spring from the seemingly dead stump—and within five years it will be twenty to thirty feet high. Kashula pointed out a hillside now covered in trees. "It's just one of our TACARE forests," he said. "Nine years ago that slope was quite bare."

The villagers gathered under the trees to greet us, including two shy scholarship girls. A ten-year-old R&S leader, confident in his tight-fitting red-striped shirt, told us about the trees his club was planting. I told them how I spoke about the TACARE villages as I traveled around the world. "And," I said, "we must remember to thank the chimps. It was because of them that I came to Tanzania—and see what it has all led to!" I ended with a chimpanzee *pant-hoot* and had all the villagers joining in.

TACARE has greatly improved the lives of the people in twenty-four villages around Gombe, generating a level of cooperation that would have been unthinkable before. And today, under the leadership of Emmanuel Mtiti, we are reaching out to many other villages in the large, mostly degraded area that we call the Greater Gombe ecosystem, with the aim of restoring the forests. Most recently, with government support, we are introducing the TACARE programs in a very large and relatively sparsely populated area to the south, hoping to protect the forests before they are cut down and thus save many of Tanzania's remaining chimpanzees.

In TACARE villages women can take out tiny loans and start their own environmentally sustainable projects—such as establishing a tree nursery. *(JGI/George Strunden)*

Emmanual Mtiti showing me, for the first time, the regenerating forest outside Gombe National Park—the leafy corridor that will enable the chimpanzees to leave the park and interact with other remnant groups. *(Richard Koburg)*

Chimpanzees, Corridors, and Coffee

The farmers in the high hills round Gombe grow some of the best coffee in Tanzania, but because of the lack of roads and transport difficulties, they often lump their superior beans with those grown at lower altitudes. Green Mountain Coffee Roasters was the first company to join us in our effort to get these farmers a good price. Now there are a few specialty brands on the market in the United States and Europe, and the farmers—as well as connoisseurs of good coffee—are overjoyed.

The goodwill generated is helping the chimpanzees as well. Every village is required, by the government, to create a Land Management Plan, which includes allocating an agreed percentage of their land for protection or restoration of forest cover. Now many of the villages are setting aside up to 20 percent of their land for forest conservation. They're also working with JGI's amazingly talented Lilian Pintea, an expert in GPS technology and satellite imagery, to ensure that these protected areas will form a corridor so that the chimpanzees will no longer be trapped in the park. That will link them to other remnant populations living in the vast habitats we are helping to protect.

Early in 2009, I stood with Emmanuel Mtiti on a high ridge looking over at the steep hills behind Gombe. A few years ago, those hills had been bare and eroded by desperate attempts to grow crops. Now I could see trees—hundreds and hundreds of them, many more than twelve feet high. This regeneration stretched as far as we could see, toward the Burundi border to the north and the town of Kigoma to the south. It was the first part of the leafy corridor about which I have been dreaming since TACARE started. A last chance for the long-term survival of the Gombe chimpanzees.

PROTECTORS OF THE WORLD OF PLANTS

For most people, mention of endangered species brings to mind giant pandas, tigers, mountain gorillas, and other such charismatic members of the animal kingdom. Seldom do we think of trees and plants in the same category—as life-forms that, in many cases, we have pushed to the brink of extinction and that desperately need our help if they are to survive.

This discussion about healing earth's scars illustrates that, through a combination of human determination, scientific know-how, and the resilience of nature, even badly compromised habitats can be restored—and time and again we find that it is plants that start the process. Somehow they take root on rock we have laid bare, on land and in water contaminated with pollutants. Slowly they build up the soil and clean the water, paving the way for other life-forms to follow.

Without plants, animals (including ourselves) cannot survive. Herbivores eat plants directly; carnivores eat creatures that have fed on plants—or, to be picky, they may eat animals that fed on animals that fed on plants.

Yet for the most part, the work of the botanists and horticulturalists who battle to save unique plant species from extinction, and to restore habitats, goes unnoticed. The more I thought about this, the more I realized that it was really important to recognize the sometimes extraordinary work that has been and is being done to preserve the rich diversity and sheer beauty of the plant life that brightens our planet. I wanted to acknowledge the contributions of the field botanists who travel to remote places to collect specimens of endangered species, the talented horticulturalists who struggle to germinate reluctant seeds, the skill and patience of those working in herbariums, seed banks, and the many Centers for Plant Conservation that have been established in so many places around the world.

Many of these scientists have generously shared their stories with me or informed me of the work of others. They are dedicated, these custodians of our botanical world. They travel to remote places, searching for rare species, collecting seeds, dangling from ropes to hand-pollinate the last individuals of an endangered plant that has taken refuge in the most inaccessible and inhospitable terrain. They have worked, year after year, to find ways of propagating, in captivity, some plant that is vanishing—or gone from the wild. Some of these heroes I have met, such as Paul Scannell and Andrew Pritchard, who have worked tirelessly for years to protect and restore some of Australia's endangered orchids, and Robert Robichaux, who has devoted his life to saving and restoring the glorious silversword and other Hawaiian plants.

Carlos Magdalena—
skilled and passionate
Kew Botanical
Gardens' horticulturist
with the café marron
he saved from
extinction.
(Carlos Magdalena)

When I visited Kew Botanical Gardens, I heard many fascinating stories about plants in the collection. Carlos Magdalena
told me about the café marron, a small flowering shrub that
was rediscovered by a schoolboy on Rodrigues Island (off
Mauritius) about forty years after it was last seen. This was
exciting, and the area was searched carefully in the hope that
other individuals would be found. It seemed, however, that
only the one plant had survived. Carlos described the nightmare of protecting it.

"It was the last specimen of a species unique in its genus,"
he said. "It did not set seed. There was no information on its
cultivation and no other similar surviving species for compari-
son. It was a few meters off a public road, on a private piece of
land, on a remote island with no botanical gardens. And frequently exposed to cyclones!"

Of course, Kew botanists wanted to save the lone survivor,
and the only way was to bring cuttings back to Kew for propagation. But the logistics of getting three hearty cuttings and
keeping them alive and safe during the many flights and long

journey from Rodriquez Island to London proved daunting. Meanwhile, now that the plant was rediscovered, it was immediately at risk of being destroyed by poachers who cut off the branches and used them for a local medicinal remedy. It took two stressful years to figure out how to get three cuttings from the sickly survivor to Carlos at Kew. And finally only one grew.

Almost two decades after its rediscovery, Carlos traveled back to the island to check on the original lone plant and he found it still alive, but stuck inside a cage that was covered with invasive weeds. "It was in poor health and attacked by two insect pests," he told me. "Even though it was in a cage and covered with weeds, somehow somebody managed to jump in and cut the plant almost to ground level." Carlos tended the plant as best he could. He got rid of the pests and the weeds, and even managed to repatriate seedlings that were growing at Kew from a clone of the surviving plant.

Carlos's seventeen-year struggle to persuade the café marron to produce fertile seeds is one of my favorite plant stories. I asked him how it had felt to be primary caretaker for a very rare specimen like the café marron.

"It is quite a responsibility," he said, "when you suspect or know for certain that if it dies in your glasshouse—the whole species goes. It has scared me to death on several occasions. Going home on a Friday in a summer heat wave and thinking: Will it be there on Monday? . . . Will the person on duty remember to water it properly? Have I watered it too much? Or too little? This is something I'm trying to get used to but I haven't yet!"

My final story is about a truly dedicated field botanist, Reid Moran. For decades he was a sort of living myth in botanical exploration in Baja California and the Pacific Islands of Mexico. In 1996, Moran wrote *The Flora of Guadalupe Island*, which describes the immense botanical richness of the island but also analyzes, with despair, the devastating impact of the goats and other introduced species. "With its unique flora it is a Mexican treasure that urgently needs protection," he said, "the most beautiful island I know . . ."

Moran retired, but one of his friends, Dr. Exequiel Ezcurra, director of the Biodiversity Research Center of the Californias

Reid Moran on Mexico's Guadalupe Island, collecting a rare specimen of the endemic Guadalupe rock daisy (*Perityle incana*) that survives on a steep bluff, out of the destructive reach of goats. Around the world, botanists like Reid have risked and devoted their lives to preserving the diversity of earth's plant species. (*San Diego Natural History Museum*)

in San Diego, was a great admirer of Moran's work. A question lingered in Exequiel's mind: Could some of this collapsing paradise, with its incredible biological richness, still be saved? An expedition was organized and it was found that the situation, overall, was bleak, with many of the island's unique species apparently gone and others seeming on the brink of extinction. Unless something was urgently done, the island would be a "paradise lost."

Exequiel told me a dramatic story about the international cooperation and heroic efforts it took to secure funding and painstakingly restore the devastated island to its glorious paradisiacal condition.

In this book, we have shared stories of islands that were restored in order to provide the right habitat for endangered animals. Guadalupe Island was restored primarily to protect its beautiful and endangered flora—although it did see the vitalization of many birds and insects.

This story illustrates, in a striking way, the resilience of nature: Many of the plants on Guadalupe Island had weathered years of a very hostile environment and somehow survived. It is truly a success story, and without the pioneering work of botanist Reid Moran it would never have happened.

And without all the other men and women who are working so hard to conserve and protect our plants and their environments, our planet would be a poorer place. Their efforts are not usually well known, yet their contributions are so important, so meaningful.

A moment of trust. When infant Flint reached out to me, my heart melted. I loved him. *(Hugo van Lawick/NGS)*

Why Save Endangered Species?

Why should we bother to save endangered species? For some, the answer is simple. My friend Shawn Gressel, of the Sioux Tribe in South Dakota, works to reintroduce the swift fox and the black-footed ferret on tribal lands. One day while we sat talking and looking at his photographs, Shawn said to me, "Some people ask me why it matters. They want to know why am I doing it. And I tell them it is because these animals *belong* on the land. They have a right to be there." He feels "obligated" to the animals he is working with.

Shawn is not alone. Many, if not most, of those I have spoken to feel much the same—even if they prefer (or have been advised) to give a scientific explanation of the importance of their work. And of course, there can be no question of the importance of protecting an ecosystem and preventing the loss of biodiversity. Yet there are millions of people who simply "don't get it." Especially if the species concerned is an insect—"Just a bug!" When the Salt Creek tiger beetle was listed as federally endangered, and federal money was released to help safeguard some of the unique and endangered habitat where it lives, there was a heated exchange of e-mails printed in the local Lincoln, Nebraska, newspaper. While many readers welcomed the decision, many others were shocked and horrified; some, too, were genuinely mystified. Here are three examples—and one hears similar opinions in many places.

A man calling himself Dick wrote, "Hundreds of thousands of species have come and gone without humans trying to save them. Even animals we killed off are probably happier now. Look at the dodo bird, what major environmental impact did all of them being wiped out have, other than sailors not having an easy lunch?"

Jill Jenkins asked, "Can someone tell me what difference it would make in our world as a whole, if this beetle were to become extinct?? I am really thankful our U.S. government wasn't around to offer grants to keep the dinosaur from becoming extinct. One half million dollars to save a bug when millions of humans are homeless and hungry. We should be ashamed!"

Then someone named J had this to say: "Now I have heard it all! I am getting so sick of our 'fine' government making kindergarten decisions like this! We need to save our humans that are inflicted with cancer and other life threatening illnesses before we care about this beetle thing! If I saw one in my house I would smash it!"

There were, of course, many letters from people who understood the importance of protecting the environment, even if they did not understand the reasons in detail. Theresa, for example, wrote: "It amazes me how spoiled-rotten Americans are, with our gas-guzzling SUVs and oversized . . . everything! If we don't nurture our habitat, our entire world will become one big Easter Island!" (The full story of the fight to save the Salt Creek tiger beetle can be found on our Web site: janegoodall hopeforanimals.com.)

It is indeed true that the expense of saving an endangered species can be exorbitant, so it is fortunate that in many countries there are laws protecting life-forms threatened with extinction. Else the damage inflicted on the natural world would be even greater. Hundreds of thousands of dollars may be spent on re-routing a road to protect the habitat of some small seemingly insignificant creature; a company may be forced to relocate a proposed development if the area is also home to an endangered species—or else buy suitable land elsewhere and even foot the bill of relocating the species concerned. (There are heartwarming accounts of all this on our Web site.) The reclamation of degraded habitats may cost us dearly, yet these efforts are among the most important facing us as we move into a new millennium.

We Need Wilderness to Nurture Our Souls

Scientists are continually providing facts and figures that can be used to explain the importance, to ourselves and our future, of preserving ecosystems. But the natural world has another value that cannot be expressed in materialistic terms. Twice a year, I spend a few days in Gombe—that's all the time I have. Of course, I hope that I will see the chimpanzees. But I also look forward to the hours I spend alone in the forest, sitting on the peak where I once sat as a young woman and looking out over the forested

valleys and the vast expanse of Lake Tanganyika. And I love to sit absorb-
ing the spiritual energy of the Kakombe waterfall as it drops eighty feet to
the rocky streambed below, the vegetation constantly moving in the wind
of falling water. No wonder the chimpanzees perform their spectacular
waterfall displays, "dancing" in the shallow water at the base of the falls,
swaying rhythmically from foot to foot, hurling huge rocks, then sitting to
watch the mystery of Water—always coming, always going, always there
in front of them. No wonder this was one of the sacred places where the
medicine men, in the old days, would come to perform their secret rituals. It is these experiences that fill my heart and mind with peace—being,
even for a short time, part of the forest, connected once more with the
mystery, feeding my soul.

Jeremy Madeiros, who has dedicated his life's work to protecting the
Bermuda petrels, or cahow, told me how he was taken to a California
redwood forest when he was eleven years old. For him, being among
those giant ancient trees was a spiritual experience, as it is for so many
of us. "It was a defining moment in my life," he told me. "It determined
my future path."

Rod Sayler, working to save the pygmy rabbit in Washington State,
believes that human values and ethics should, where possible, drive the
saving of endangered species. "We are treading too harshly on the earth
and consuming and degrading too much of the planet," he said. "If we
allow extinctions to happen through ignorance or greed, then with the
loss of each endangered species and unique population, our world becomes less diverse and strikingly less beautiful and mysterious. Our
oceans, grasslands, and forests will echo with silence, and the human
heart will know that something is missing—but it will be too late." He
argues that, although the fight to save endangered species may be costly,
"can the human spirit afford not to try? If we do not, someday we will
look back with the wisdom of time and regret our decision."

The Keepers of the Planet: What Keeps Them Going

Fortunately for the future of the planet and all its life-forms, including
us and our children, there are, as we have seen, brave souls out there
fighting day after day to save what is left and restore what has gone.
Working on this book has been a real privilege, for I have met so many
of these extraordinary, dedicated, and passionate people from around
the globe. Many of them, as described, have spent years working in remote places, enduring considerable personal discomfort and sometimes

very real dangers. They have had to battle, too, not only with the harsher aspects of nature but also with uninformed, unimaginative, and short-sighted officials who refuse permission to move ahead with urgently needed management actions. Yet they have not given up.

What keeps them going? I asked some of those who have been longest in the field. All of them confessed to loving the wilderness, being out there with nature. And, as well, they became utterly absorbed in the work—almost, for some of them, it was like a mission. They simply couldn't give up. They became, as the wife of Dean Biggins (one of the black-footed ferret team) put it, "obsessed."

Don Merton has devoted his life to saving endangered birds. This is Adler, a juvenile kakapo—one of the many island birds that Don heroically rescued from the brink. "If you didn't love and respect the creatures you are trying to save you couldn't spend decades crawling through treacherous terrain and dangling from ropes," he told me. (The dramatic story of saving the kakapo, the only flightless parrot in the world, is told on our Web site.) *(Margaret Shepard)*

Don Merton, who has worked so hard to protect island birds, told me that most of all he loved "the ultimate challenge—fighting to save the last few individuals of a unique life-form. The black robin is one of New Zealand's living treasures . . . I felt a massive responsibility to current and future generations to save this fantastic little bird from the brink of extinction." He told me that he could hardly wait to get back to the field each spring to find out how the individual birds had fared. And, he said, "Some of my colleagues became annoyed with me when I rose very early to start searching at first light, and woke them!"

Chris Lucash, after twenty-one years with the Red Wolf Recovery Program, told me that during the early years when they

Graduate student Len Zeoli with a highly endangered Columbia Basin pygmy rabbit. "How can you see one, know one, and not love these little creatures," he said to me. "That's what drives us. That is what keeps us going." (Dr. Rod Sayler)

were releasing the wolves into the wild, he felt privileged to have the opportunity of being part of something he believed was so very important. "I had unwavering energy," he said. "I had a difficult time sleeping and wanted only to stay out keeping track of the wolves and try to figure out everywhere they went, what they did, why they did it, what they ate. I took little or no time off. I lived red wolves, and was baffled, confused by, and almost intolerant of people—friends and family—who did not feel the way I did about the program." And still, after more than twenty years working with the wolves, he looks forward to going to work "every day—sometimes even on Sundays!"

Daring to Admit That We Love

There is another aspect of their work that for some may be the most important—the relationship they establish with the animals they work with. I
have described my own feelings for so many of Gombe's chimpanzees. The
one I loved best was David Greybeard, the first who lost his fear of me, who
allowed me to groom him and tolerated my following him in the forest. And
I remember, as though it happened but yesterday, the day I offered him a
palm nut on my outstretched hand. Not wanting it, he turned away, but
then he turned back and, looking directly into my eyes, took the nut,
dropped it, then very gently squeezed my hand with his fingers. A chimpanzee gesture of reassurance. And so we communicated perfectly, he and
I, with shared gestures that, surely, predate our human spoken language.

Unfortunately in our materialistic world, where all that counts is the
bottom line, human values of love and compassion are too often suppressed. To admit you care about animals, that you feel passionately
about them, that you love them, is sometimes counterproductive for
those in conservation work and science. Emotional involvement with
one's subject is considered inappropriate by many scientists; scientific
observations should be objective. Anyone who admits to truly caring
about, having empathy with, an animal is liable to be written off as sentimental, and their research will be suspect.

Fortunately, most of the extraordinary individuals whose work is discussed in this book are not afraid of showing that they care. (Particularly
those who have retired!) During one of my discussions with Carl Jones, of
Mauritius Island fame, he echoed my own belief—that although scientists
must have the ability to stand back and observe objectively, "they should
also have empathy." Humans, he said, "are intuitive and empathetic before they are coldly scientific"—and he believes that most "scientists call
on these underlying qualities every day." When he was working to save the
Mauritius kestrels, he got to know and understand each bird as an individual. Don Merton waxed lyrical over the black robins, "those delightful,
tame, friendly little birds." Over the years, Don said, "I naturally became
very attached—even emotionally involved you might say! I just loved
them." And Len Zeoli, when I asked him what motivated him to keep on
working to save the pygmy rabbits, said simply, "How can you see one,
know one, and not love these little creatures? That's what drives us. That
is what keeps us going."

Mike Pandey, while filming in India the barbaric method of killing
gentle, harmless whale sharks, came across a huge individual who was

dying. "It slowly turned to look at me . . . beseeching and pleading . . . the intelligent eyes spoke a million words." He said he would never forget that look: "Suddenly I was in communication with the majestic creature and there was a deep-rooted bonding." That was the turning point that transformed his life. He decided to "speak out for the voiceless" and started his long series of powerful films for conservation.

Brent Houston told me of the time when a young black-footed ferret approached him as he sat near the den, in the first light of day. "Without warning, he approached my foot and sniffed my hiking boot . . . I thought the pounding of my heart would scare him, but I remained still, desperate for some sort of connection. He looked right up at me and at my face, into my eyes. And then the most extraordinary thing happened. This young ferret, looking up at me with his big round eyes, put his little black foot on my hiking boot and he held it there. I looked right at him and he looked at me and he saw me smile. It was one of the most satisfying moments in my long career of observing wildlife. Here was one of the last black-footed ferrets in the world reaching out to me, trusting me, perhaps even asking for my help."

It is this—this link between the human being and the other animals with whom we share Planet Earth, this connection we can establish with another life-form—that for many makes it possible to carry on. To carry on with work that can be so hard, carry on despite the frustrations and setbacks, and sometimes the outright hostility or ridicule of those who believe that to save any species from extinction is sentimental and a waste of money and resources.

But they cannot do it alone, these Keepers of the Planet. To save Planet Earth, each of us who cares must become involved in protecting and restoring the wild places and the animals and plants that live there. We hope that this book, together with our Web site, overflowing with stories of passionate, dedicated, and always hopeful people, whose efforts have saved myriad life-forms from extinction, will encourage those who are out there now, working tirelessly as they try to save other highly endangered animals and plants, each one precious and unique. And those who are striving to prevent further species from becoming endangered. And yet others fighting to restore and protect the environment. Their tasks sometimes seem almost impossible—and if they had no hope of success, they would surely give up.

If we are without hope we fall into apathy. Without hope nothing will change. That is why we feel it is so desperately important to share our own, irrepressible hope for the animals and their world.

Don Merton saying good-bye to kakapo Richard Henry, April 2010. Richard Henry was the last known survivor of his species from mainland New Zealand and was believed to be around 100 years old when this photo was taken. *(Errol Nye)*

Epilogue

It's been almost two years since the original release of *Hope for Animals and Their World*. And once again, I am writing from my childhood home in Bournemouth during a brief seven-day respite from my rigorous lecture tour schedule.

Editions of this book have now been published in the United Kingdom and Australia, and it has been translated and published in China, Germany, Hungary, the Netherlands, and South Korea. And there are ongoing negotiations with publishers in other countries. What makes me happy is that, as I travel around the world, so many people tell me how they have been inspired by the hopeful nature of the stories and given new energy to carry on fighting to conserve the natural world. This, of course, is why I wrote it. In particular I am thrilled that a number of teachers are using the book, and it is finding its way into school libraries. In fact, several Roots & Shoots groups have been formed by students who read the book and became inspired to do their bit to help change the world.

Of course, inevitably, there are those who have criticized me for burying my head in the sand, ignoring the facts concerning rate of forest loss, ocean acidification, species loss, pollution, industrialization—and all the other threats to the natural world and the life forms that inhabit it. I am not, and I have stated that several times throughout the book. The overall picture is grim. The recent oil spill in the Gulf is a shocking reminder of the extent of the sacrifice demanded by those who would put the bottom line ahead of concern for the generations (of all life forms, including our own) to come.

There have been many other suffocating spills; there will be others. Human avarice knows no bounds and the economic situation provides the perfect excuse for cutting corners in the interests of quick profit. Chernobyl is not the only nuclear fuel disaster. And let us not forget the Bhopal gas tragedy and all the other environmental and humanitarian nightmares caused by industrial toxins.

I think about these things all the time. I agonize over the destruction and the suffering. It would be so easy to give up, to bewail human frailty, and to retire to the house and garden in Bournemouth.

But I think, too, about the wonderful people I have met, and about whom I have written in this book. I think about the marvelous species, which, thanks to them, have been given a second chance. I think about the ecosystems that have been restored, against all odds. And, above all, I think about the young people, their enthusiasm, their determination to save the world. There is something about the young people today. They have wisdom beyond their years, they are courageous. They know the stakes—sometimes only too well. If they give up, then indeed there is no hope for any of us, man nor beast nor plant.

And so I chose to try to balance the negative with the positive, to share inspirational, hopeful stories to balance the gloomy and hopeless information with which we are bombarded every day in our newspapers, on television, on the Internet. And I find there is growing a new spirit, a new determination, a new desire in people to do their bit, to make the right choices, to let their voices be heard.

Oh yes, I know how bad things are. I know that if there is not a major change nothing can save our planet. But I also know that there is still a chance we can turn things around. But only if we have hope. The people about whom we have written in this book have not given up, and nor shall I.

Sadly, we lost three of our heroes since the book's original publication. Devra Kleiman, who worked tirelessly and successfully for forty years to conserve and protect the golden lion tamarin and also helped with the breeding program for the giant panda in China, died from cancer in April 2010. Before she passed away she proposed setting up the Devra Kleiman Fund to Save the Golden Lion Tamarin. She wanted all the contributions to support the conservation work for golden lion tamarins in Brazil. This will help the program to continue into the future. To make a contribution go to www.savetheliontamarin.org.

Ernie Kuyt died in May 2010 from injuries he sustained in a fall. Ernie was one of the key players in the whooping crane story for more than twenty-five years. Even after he retired, he remained deeply committed to the whooping cranes. I am so glad I met Ernie and his wife in the spring of 2009 when I was giving a talk in Edmonton, Alberta. Ernie and his wife came for tea and we spent a good hour talking about cranes and his hopes for their future. What a wonderful man. The world is a poorer place without him.

John Thorbjarnason, whose work to conserve the Chinese alligator is presented on our website, and is included in the Chinese edition of the book, also died in 2010. After surviving numerous close encounters with alligators and crocodiles in the wild, John succumbed to a particularly lethal form of malaria while studying the dwarf crocodile in Uganda.

In the fall of 2010 I learned that my good friend Don Merton had been diagnosed with cancer. Don is one of the most successful and re-spected heroes in the conservation world. He dedicated his adult life to saving endangered species and was responsible for saving the black robin from extinction. He is a true pioneer who developed many of the island restoration and species translocation techniques that are used throughout the world today. And he worked the hardest and with the most success in the efforts to save the New Zealand flightless parrot, the kakapo, from the brink of extinction. The full story is on our Web site (janegoodallhopeforanimals.com) and is featured in the Australian edition of the book.

He wrote me a beautiful letter in the fall of 2010, telling me I should not be sad and that he had had "a great life, worked with many amazing, dedicated folk, and visited some of the most remote and beautiful places imaginable."

Of course, I had tears running down my cheeks even before I read the next paragraph of his letter:

"One of the many things I wanted to do when I heard the news was to say good-bye to old Richard Henry kakapo. So my son Dave made ar-rangements for he, my wife Margaret, and I to make a quick trip to Codfish Island to do just that. RH is clearly very old and frail now—just a shadow of his former self—and but for ongoing supplementary feeding would not be alive…his weight is OK (2.2kg), but he is blind in one eye and appears not to have much sight in the other and his plumage is not nearly as bright as it used to be. We had a wonderful few days on Cod-fish—fine and warm, tracks dry and being spring time here, the forest birds were all singing well."

The picture I had in my mind, after reading that letter, of Don revisiting his friend Richard Henry, whom I felt I knew so well after writing his story, is so beautiful and so poignant.

Then, in the beginning of 2011, just as I was finishing this Epilogue, we received an e-mail from Don telling us that Richard Henry was found dead on December 24, 2010. He was believed to be about one hundred years old when he died. Don told me that there are now 121 kakapo alive—the first of the original females caught in the 1980s on Stewart Island died just a few months before Richard Henry. In the past few years the total population has increased by thirty-five individuals, but the species suffers from a lack of genetic diversity and so the breeding success of Richard Henry's progeny—two males and one female—will be very important for the future of the kakapo.

Since this book was published we have kept in touch with our other heroes, and regularly posted updates from their projects on our Web site. Most of the news is positive, but one really tragic story was sent in by Frank Zino. He wrote that last year (2010) an arsonist started a fire on Madeira that destroyed the island's nesting grounds for the Zino's petrel.

"From our prospective a whole generation was wiped out," Frank wrote. Most of the parents were at sea when this happened, which means more pairs will breed in the new nests that were hastily built for them, and more chicks will come. But it is heartbreaking for all of us when a generation of these fragile creatures is devastated.

Yet Frank and his team have no thought of giving up. They are working to improve the damaged nest sites and create others, making new plans to help ensure the future of the species they have already worked so hard to save.

There is also bad news about the Sumatran rhino—the population has decreased by one hundred individuals. But there is a silver lining—the United States has agreed to cut the Indonesian debt by nearly $30 million over eight years in exchange for increased protection of Sumatran forests that are home to these endangered rhinos (as well as tigers and orangutans). This debt-swap-for-nature deal, orchestrated by Conservation International and the Indonesian Biodiversity Foundation, creates a trust to preserve 18.29 million acres, including Way Kambas National Park. It is the largest of fifteen debt-for-nature swaps that have been approved since the U.S. Tropical Forest Conservation Act was passed in 1998.

As I said, most of the updates are positive. I heard from Travis Livieri that black-footed ferrets have been successfully released into their old stamping grounds on the Canadian prairie, in the Grasslands National Park in Saskatchewan. A film was made, *Return of the Prairie Bandit*— which was released in February 2011. A longer version is being made for showing at wildlife film festivals. Travis also has a new footprint of Mom, made of fiberglass. Mine was plaster of Paris, and was beginning to chip around the edges.

The fall 2010 migration of the Waldrapp was, says Fritz Johannes, "fantastic and extraordinary." For the first time the flight distance and speed per day was comparable with that of wild birds. All fourteen birds performed perfectly, covering a distance three times longer than previous years, as it had been decided to lead them around instead of over the Alps. They are now in their winter home—and everyone is waiting to see how they make out on the return flight to Austria in the spring. Check out our Web site for news.

And here are some other updates. Since this book was first published, the wild population of Vancouver Island marmots has increased by over 120 individuals. There are 150 more wild crested ibis in China. The number of nesting pairs of cahow has increased by at least twelve. And it seems that the American burying beetle has been successfully re-established an on Nantucket Island—the population is remaining stable and most likely growing.

I just received some news from John Hare about his wild camels. For years there has been a controversy as to whether they were descended from Silk Road runaways—from feral Bactrians—or if they were a quite separate species. Finally, in 2010 this question has been settled: The wild Bactrian camel has been officially declared an entirely separate species, one that has been around for 700,000 years. There wasn't time to rewrite the chapter in this book before it was reissued in paperback, so we have left it with the original name and information. But from now on it will no longer be referred to as *Camelus bactrianus ferus*, the Wild Bactrian Camel, but simply *Camelus ferus*, the Wild Camel.

Finally there is an exciting story about the short-tailed albatross that I want to share here. Hiroshi Hasegawa assures us that not only is the population on Torishima still increasing in numbers, but the new Japanese breeding grounds are working out really well with many nests, and many chicks fledging.

But what is truly wonderful is that in December 2010 two pairs of the short-tailed albatross were observed nesting in small islands of the Hawaiian archipelago. For years, seabird workers had been using decoys to try and lure mating pairs to nest in the archipelago. Then finally, a nest was found on Eastern Island, one of the islands of Midway Atoll National Wildlife Refuge; and another nest fifty-five miles away on the small island of Kure Atoll. Until these nests were discovered, the only remaining breeding colonies of the birds were thought to have been on the Japanese islands. Then, in January 2011 came the news that the Eastern Island pair successfully hatched a chick. This is the first time the species has ever been known to successfully breed outside Japanese waters!

And so, amidst the successes and the setbacks, the work continues— one far-off nest, one chick, one life, one species at a time. More than anything it is the dedication of so many humans that continues to sustain my hope. My hope for animals. My hope for this world.

—Jane Goodall, February 2011

Historical photo taken on my veranda in Dar es Salaam on the day Roots & Shoots began in February 1991. *(JGI)*

Appendix
What You Can Do

I meet so many people as I travel around the world who are deeply depressed by what is happening to our planet. The media are continually publishing, among a great deal of other shocking news, stories of deadly pollution, melting ice caps, devastated landscapes, loss of species, shrinking water supplies, and all the rest. In the face of such desperate information—which unfortunately is mostly true—people tend to feel helpless and often hopeless. "How can you remain optimistic?" is, as I have said, the question I am asked most often.

The best way I know to counteract despair is to do everything I can to make a difference, even in the smallest way, every day. To take some action to do *something* about at least some of the bad things that are going on. That is why I left Gombe and the forests I love—to try to do my bit to raise awareness of the plight of the chimpanzees and their forests, and to do whatever I could myself.

It is also important to realize that bad news is more likely to be published as being more "newsworthy." In fact, there are also many truly wonderful things going on as people work selflessly to make this a better world. One of the reasons we wanted to write *Hope for Animals and Their World* was to share some of the good news.

Throughout this book and on our Web site, there are stories of biologists who are working tirelessly to save endangered species. But there are countless others, members of the "general public," who also play a vital role. They often get no credit, their names, outside the area where they live, often unknown. And because sometimes their actions—demonstrating against some destructive plan of industry or government or writing letters to the relevant authorities—are not always successful, the true significance of the role they play is often underestimated. Yet in the long run, these are the people who truly matter. They donate their money, skills, or time, they help to raise awareness and persuade others to join them.

In all walks of life, people are contributing to the growing awareness of what is going on—writers, photographers, filmmakers, and those guiding an increasingly eager public on trips into nature. NGOs, with their education programs, encourage people to volunteer in field projects, learning about the natural world and taking action to help protect it. Landowners may sign Safe Harbor agreements, protecting the habitat of an endangered species; others may sign a conservation easement, receiving financial benefits for helping wildlife by *not* developing or cultivating their land.

And then there is the role played by youth. Why am I devoting so much time to working with children? Because it is not much use for me or anyone else to work desperately to save animals and their world if we are not, at the same time, educating our youth to be better stewards than we have been.

ROOTS & SHOOTS—WHAT YOUTH CAN DO

In view of the gloom and doom everywhere, I was hardly surprised to find, as I traveled around the world, that so many young people seemed depressed, angry, or apathetic. It was, they told me, because their future had been compromised and there was nothing they could do about it.

We have indeed compromised their future. There is an indigenous proverb: "We have not inherited this planet from our parents; we have borrowed it from our children." But it is not true. When you *borrow*, there is the intention of paying back. We have been relentlessly *stealing* our children's future. Yet it is not true there is nothing to be done about it.

My contribution was to start our Roots & Shoots humanitarian and environmental program. This encourages its members to roll up their sleeves and undertake projects that improve things for people, for animals, and for the environment—projects that have a positive impact on the world around us. Its most important message is that every individual matters and has a role to play—that each of us makes a difference *every day*. And that the cumulative result of thousands of millions of even small efforts is major change.

The name is symbolic. The first tiny roots and shoots of a germinating

Roots & Shoots youth leader Washo Shadowhawk, from Beaverton, Oregon, has been passionate about saving injured animals, especially snakes, since age two. Shown here with Sandy the bearded dragon and Monty the python, whom he rescued from the exotic pet trade. *(Meadow Shadowhawk)*

seed look so tiny and fragile—hard to believe this can grow into a big tree. Yet there is so much life force in that seed that the roots can work their way through boulders to reach water, and the shoot can work its way through cracks in a brick wall to reach the sun. Eventually the boulders and the wall—all the harm, environmental and social, that has resulted from our greed, cruelty, and lack of understanding—will be pushed aside. Just as hundreds and thousands of roots and shoots—the youth of the world—can solve many of the problems their elders have created for them.

The program was born in Dar es Salaam in 1991 when twelve students, representatives from several secondary schools, had gathered on my veranda to learn more about the behavior of Tanzania's wildlife. They were shocked to hear about poaching and other problems and wanted to learn more and help. So they started clubs in their schools, and we organized gatherings to discuss such issues.

How amazing that from such a simple beginning the program has, at the start of 2009, spread to about a hundred countries with some nine thousand active groups, involving young people from preschool through university and beyond. Roots & Shoots is unique in many ways: It links young people from many different cultures, religions and countries; combines care and concern for animals, people, and the environment; and involves people of all ages—there are even groups in retirement homes and prisons! With its shared philosophy, it is spreading seeds of global peace. And it is creating leaders of tomorrow's world who understand that this life is about more than just making money.

I encourage you to check out the book's Web site (janegoodallhopeforanimals.com) where I have brought together information about the myriad wonderful projects undertaken by Roots & Shoots to benefit wildlife. And it profiles some quite exceptional young people who are part of our Roots & Shoots Global Youth Leadership Council. I hope that, whatever your age, you will become involved in some way with this program or one of the thousands of other youth groups around the world. You can be a part of this movement simply by consciously doing your bit, every day, to make this a better world for all life. That is the best antidote to despair that I know.

This section is for everyone, young or old, who cares about the animals with whom we share the planet and who is tired of sitting on the sidelines. It provides information about ways you might help the species profiled in this book, organizations you can contact, ways in which you might volunteer.

A CRITICAL MASS

But what is most important is that you do *something*. Don't feel that, because you cannot do all you would like to do (if you only had more time, more money, more influence), then it is better to do nothing. When you read, in your local paper, that a woodland area you love is slated to be developed, don't just sigh and shrug—take action. Any action. Find out more, who is involved, why it is happening. Write letters. Attend meetings in your town hall. Make your views known. You may not succeed—but you may. If you don't try, you won't know.

If one of the stories in this book captures your imagination, moves you and you want to help—contact the relevant organization and ask what you can do. And remember, even if you can only afford to send a small donation—it was thousands of tiny amounts donated that made President Obama's election campaign so hugely successful!

It is going to take a critical mass of people who truly care about the future of our planet and our children to turn things around. Please join

WHAT YOU CAN DO

forces with the remarkable, dedicated men and woman whose efforts are described in these pages. Please help us to achieve our goal of saving the animals and their world.

GLOBAL ACTION

For many of the endangered animal and plant species in this book, the following resources may help you obtain more information about conservation activities and learning how you can make a difference.

- Support the Jane Goodall Institute (JGI). The organization advances the power of individuals to improve the environment for all living things. While continuing Dr. Jane Goodall's efforts to study and protect chimpanzees, JGI has also become a leader in innovative conservation approaches that better the lives of local people. In addition, the institute's global youth program inspires young people of all ages to become environmental and humanitarian leaders. Visit JGI's Web site: www.janegoodall.org.
- Join Jane Goodall's Roots & Shoots. The Roots & Shoots global network connects youth of all ages who share a desire to create a better world for people, animals, and the environment. Hundreds of thousands of young people all over the world identify problems in their communities and take action through service projects, youth-led campaigns, and an interactive Web site. To take part, please visit www.rootsandshoots.org.
- Help communities, protect chimpanzees. JGI enables communities living near chimpanzee habitat to become partners in protecting these amazing creatures by assisting villagers with their most immediate needs such as water, sanitation, and health care, while fostering livelihoods that do not harm the environment. Visit JaneGoodall.org to learn how you can donate to JGI and support these critical programs.
- Become a Chimpanzee Guardian. At the Tchimpounga Chimpanzee Rehabilitation Center in the Republic of Congo, JGI provides safe and caring habitats for orphan chimpanzees—victims of the illegal commercial bushmeat and pet trades. You can support this effort by becoming a Chimpanzee Guardian. To find out more, please visit www.janegoodall.org/chimp_guardian.
- Support Durrell Wildlife Conservation Trust. This organization is directly involved with the conservation and protection of many species in this book. Visit www.durrellwildlife.org to read more, to adopt animals, and to make donations.

- Play a role in conservation by contacting the Nature Conservancy at www.nature.org. This organization protects a variety of vital marine and land habitats. You can get involved by becoming a member, subscribing to the e-newsletter, making donations, and even creating your own personalized nature home page.
- Get inspired through Conservation International at www.conservation.org. This nonprofit protects threatened species and habitats through an innovative approach that combines the best of what community and science each has to offer. In addition to making donations, you can take action by calculating your impact on the earth, obtain information about ecotourism, support individual campaigns, and learn about career opportunities in conservation.
- Join the World Wildlife Fund at www.worldwildlife.org to work toward a stable, sustainable future for humans, animals, and ecosystems. You can make donations, adopt an animal, become a member of the conservation action network, or support individual campaigns such as Take Action for Earth Hour.
- Protect wildlife and wild habitats throughout the world by contacting the Wildlife Conservation Society at www.wcs.org. In addition to its global conservation projects, this organization also manages several wildlife parks in New York such as the Bronx Zoo and the Central Park Zoo. You can get involved by visiting one of the parks, becoming a member, donating your time or funds, and supporting individual campaigns such as the No Child Left Inside Act.
- Be a part of the solution to some of the world's most complex environmental challenges by contacting the International Union for the Conservation of Nature at www.iucn.org. You can learn more about their global programs, search through their database for a conservation organization near you, and become a donor.
- Check out www.fieldtripearth.org—a project of the North Carolina Zoological Society. Field Trip Earth is a global resource for teachers, students, and proponents of wildlife conservation.
- Enroll in an earth expeditions course at Earth Expeditions (www.earthexpeditions.org) or earn a master's degree through the local field program (www.projectdragonfly.org). These programs engage teachers, environmental professionals, and others directly in conservation at field sites worldwide.
- Educate yourself about the world's most endangered species by visiting Earth's Endangered Creatures at www.earthsendangered.com.

This Web site lists the facts about each species as well as organizations you can contact that are involved in their conservation.

- Contact your favorite conservation organization such as the National Audubon Society at www.audubon.org, Defenders of Wildlife at www .defenders.org, National Wildlife Federation at www.nwf.org, or the Environmental Defense Fund at www.edf.org to get involved or make a donation.
- Listen to howls, screeches, roars, and much more whenever you receive a phone call. The Center for Biological Diversity offers free endangered species ringtones and phone wallpapers. Visit www .rareearthtones.org/ringtones to learn more. You can contact the center directly at www.biologicaldiversity.org to find out more about their conservation campaigns and offer your support.

DOS AND DON'TS

Below is a list of dos and don'ts to help protect the animal, plants, and habitats of our fragile Planet Earth.

- Support only those zoos and aquariums that are accredited with the Association of Zoos and Aquariums (www.aza.org).
- Drive safely, since many animals must cross roadways in order to find food.
- Keep the roads clean. Litter attracts wild animals, causing them to be hit by cars.
- Be aware of what's happening on your public lands (national parks, national forests, Bureau of Land Management) and how they are managing wildlife.
- Don't purchase products from companies with bad environmental records. Your financial support will only encourage bad behavior.
- Be mindful of plant and animal habitat when you are out in nature, on bike trails, and so forth.
- Support organizations that advocate for sustainable ocean policies, particularly in international waters.
- Limit seafood intake and educate yourself by visiting the Monterey Bay Aquarium's Seafood Watch Program at www.montereybay aquarium.org. Print its pocket-size recommendations or download them onto your i-phone.
- Reduce your carbon footprint. Find ideas and inspiration from the Global Footprint Network at www.footprintnetwork.org.

ANIMALS AND INSECTS

Abbott's Booby

Take Action

- Learn more about the Christmas Island Seabird Project at www .biologie.uni-hamburg.de/zim/oeko/seabird_e.html. Sponsored by the University of Hamburg, this project investigates the Abbott's booby as well as two other threatened bird species—the Christmas Island frigatebird and the red-tailed tropicbird—in order to improve conservation efforts. There are opportunities for students to get involved through field assistantships and research projects.

Meet the Species

- Visit Christmas Island National Park and take a guided bird-watching tour. For more information about the park, contact the Christmas Island Tourism Association at www.christmas.net.au.

American Burying Beetle

Take Action

- Contact the Roger Williams Park Zoo located in Providence, Rhode Island, at www. rwpzoo.org. You can learn more about their award-winning American burying beetle recovery program as well as make donations to the beetle and other endangered species.
- Remove electronic bugzappers. These attract and kill burying beetles in addition to many other beneficial insect species.

Meet the Species

- Visit a zoo that has a captive breeding program such as Roger Williams Park Zoo and the Saint Louis Zoo in Missouri.
- Watch a video of recently hatched beetle grubs through the Saint Louis Zoo Web site at www.stlzoo.org.

American Crocodile

Take Action

- Visit the Croc Docs, a University of Florida–sponsored Web site at http://crocdoc.ifas.ufl.edu/index.htm. Here you can learn more about animal behavior and conservation efforts, view research publications, and take virtual field trips to Belize and southern Florida to observe crocodiles.
- Never try to feed or entice crocodiles. Not only is it illegal and dangerous, but it may encourage them to approach humans for food in the future.

- Always dispose of fish scraps in a garbage can. Do not throw them in the water.

Meet the Species

- Visit the species at Everglades National Park, Biscayne National Park, or Ding Darling National Wildlife Refuge.
- See the American crocodile at the Philadelphia Zoo, Pennsylvania, or the Central Florida Zoo, Florida.

ANGONOKA (PLOUGHSHARE TORTOISE)

Take Action

- Contact the Turtle Conservation Fund at www.turtleconservation fund.org to learn more about the latest projects this organization is pursuing to ensure the long-term survival of the ploughshare as well as other highly endangered tortoises, and to make donations.
- Assist in conservation efforts by contacting Durrell Wildlife Conservation Trust at www.durrellwildlife.org. In addition to monitoring and studying the tortoises in what remains of their wild habitat, Durrell is supporting local communities surrounding Baly Bay to protect ploughshares and their habitat. By visiting their Web site, you can make donations, participate in an adoption program, and search through a list of ecotourism opportunities including a trip to Madagascar.

Meet the Species

- Take a field trip to the Honolulu Zoo, Hawaii, to visit the ploughshare tortoise.

ASIAN VULTURES

Take Action

- Contact the Royal Society for the Protection of Birds at www.rspb.org.uk and the International Centre for Birds of Prey at www.icbp.org. Here you can learn more about conservation efforts and make donations to the organizations' captive breeding programs.
- Contact the Peregrine Fund: World Center for Birds of Prey at www.peregrinefund.org. You can learn more about the Asian Vulture Population Project, as well as assist in the effort by providing information on any vulture colonies you may be aware of in the wild.
- Educate others about the dangers of diclofenac—an anti-inflammatory drug used on domestic livestock. Vultures are killed by feeding on a dead animal that has been treated with this drug.

ATTWATER'S PRAIRIE CHICKEN

Take Action

- Contribute to the Adopt-a-Prairie-Chicken Program. Send funds that contribute directly to raising APCs in the wild to:
  Adopt-a-Prairie-Chicken Program
  Texas Parks and Wildlife Department
  4200 Smith School Road
  Austin, TX 78744

Meet the Species

- Take a trip to the Attwater Praire Chicken National Wildlife Refuge in Eagle Lake, Texas. In addition to plenty of bird-watching, there are volunteer opportunities as well. For more information, visit www .fws.gov/southwest/refuges/texas/attwater.

BACTRIAN CAMEL

Take Action

- Contact the Wild Camel Protection Foundation at www.wildcamels .com. You can learn more about current conservation efforts, become a member of the foundation, and sponsor a camel.

Meet the Species

- Visit Bactrian camels at an AZA-accredited zoo, such as the Denver Zoo, Colorado, or the Los Angeles Zoo & Botanical Gardens.

BERMUDA PETREL (CAHOW)

Take Action

- Contact the Bermuda Audubon Society at www.audubon.bm to learn more about conservation efforts, obtain information about Bermuda bird-watching tours, and make donations to cahow recovery programs.
- Visit the Bermuda Aquarium, Museum and Zoo at www.bamz.org to make donations for the recovery of cahow and other endemic species, to learn more about the Bermuda Biodiversity Project, and to donate your time as a volunteer.

Meet the Species

- Contact the Bermuda Institute of Ocean Sciences at www.bios.edu to inquire about ecotours or educational opportunities.

Black Robin (Chatham Island Robin)

Take Action

- Visit the Royal Forest & Bird Protection Society of New Zealand at www.forestandbird.org.nz to make general donations, to learn about volunteer opportunities, and to join this organization.
- Send donations for conservation efforts to:
  New Zealand Threatened Species Trust
  c/o Royal Forest & Bird Protection Society of New Zealand Inc.
  PO Box 631
  Wellington, New Zealand

Black-Footed Ferret

Take Action

- Contact Prairie Wildlife Research at www.prairiewildlife.org. Through this organization, you can learn more about the species, sponsor a ferret through their adoption program, or donate directly to on-the-ground conservation efforts.
- Support organizations that work to conserve prairie dogs (black-footed ferret habitat) such as the Prairie Dog Coalition and the Nature Conservancy.
- Explore the prairie and learn about it. It is North America's most endangered ecosystem because people simply take it for granted.

Meet the Species

- Take a guided night walk at Wind Cave National Park in South Dakota during late summer. Do not go out spotlighting alone. It is illegal without a permit and could disrupt the work of biologists or disturb the animals.
- Visit a zoo that breeds black-footed ferrets in captivity such as Louisville Zoo, Kentucky; Toronto Zoo, Ontario; Phoenix Zoo, Arizona; National Zoo, Washington, DC; Cheyenne Mountain Zoo, Colorado.

Blue-and-Gold Macaw

Take Action

- Visit the Carl H. Lindner Jr. Center for Conservation and Research of Endangered Wildlife (CREW) at www.cincinnatizoo.org. This state-of-the-art research facility is dedicated to saving endangered plant and animal species such as the blue-and-yellow macaw from extinction.

- Contact Bernadette Plair, director of the Center for the Rescue of Endangered Species of Trinidad and Tobago (CRESTT), at bplair-crestt@gmail.com if you are interested in contributing to conservation efforts. You can donate to this organization as well by mailing checks to:
  Center for the Rescue of Endangered Species of
      Trinidad & Tobago
  Attn: Alex deVerteuil
  119 Roberts Street, PO Box 919
  Port-of-Spain, Trinidad
- Do not destroy the habitat of the birds.
- Do not poach the nests for chicks.
- Do not buy macaw pets that are not captive-bred.

CALIFORNIA CONDOR

Take Action

- Visit the California Condor Recovery Program at www.cacondor conservation.org to find out more about conservation efforts, to make donations, to volunteer as a nest monitor, and to subscribe to the newsletter. This Web site also offers ideas for classroom activities to teach students about these endangered birds and what they can do to help. For more specific information about becoming a nest monitor, contact Estelle Sandhaus at esandhaus@sbzoo.org or Joseph Brandt at hoppermountain@fws.gov.
- Volunteer to take part in condor habitat enhancement on Hopper Mountain National Wildlife Refuge Complex lands and other public lands around the Central and Southern California region. To get involved, contact the US Fish and Wildlife Service at www.fws.gov.
- Make donations to the Condor Survival Fund, which supports worthwhile projects not otherwise covered by funding from other organizations or governmental agencies involved in the condor program. To make a tax-deductible contribution, send checks payable to the Condor Survival Fund to:
  Office of Accounting & Human Services
  Santa Barbara Museum of Natural History
  2559 Puesta del Sol Road
  Santa Barbara, CA 93105
- Never feed or approach a condor.
- Don't leave garbage or poisons such as antifreeze in the wild.
- If you are a hunter: Use nonlead bullets, such as copper. Be sure of your shot, and never leave any of your game behind. Bury the gut pile

of field-dressed game to make it less obvious to scavengers. Don't hesitate to report any illegal shootings.

Meet the Species
- Visit California condors at the San Diego Zoo.

CASPIAN HORSE

Take Action
- Contact the Caspian Horse Society of the Americas at www.caspian .org, or the Caspian Horse Society (UK) at www.caspianhorsesociety .org.uk, to obtain more information about the breed and to become a supporting member of these organizations.

Meet the Species
- Visit the Caspian horse at the Memphis Zoo, Tennessee.

COELACANTH

Take Action
- Learn more about the species by visiting the South African Institute for Aquatic Biodiversity (SAIAB at www.saiab.ac.za), a worldwide resource for coelacanth information and an internationally recognized center for the study of aquatic biodiversity.
- Limit your seafood, and if you do eat seafood, eat conscientiously. Visit the World Wildlife Fund's Southern African Sustainable Seafood Initiative (SASSI) at www.wwfsassi.co.za. The program educates wholesalers and restaurateurs, as well the general public, about buying sustainable seafood in South Africa and avoiding exploited or endangered sea animals.

COLUMBIA BASIN PYGMY RABBIT

Take Action
- Contact the Nature Conservancy, a nonprofit organization that has protected more than thirty thousand acres of shrub-steppe habit for the pygmy rabbit and several other species that make this area of Washington State their home. Go to www.nature.org to learn more about conservation efforts, to become a volunteer, and to make monetary or real estate donations.
- Work directly with the reintroduction and research of pygmy rabbits by contacting the endangered species program of Washington State University and the Washington Department of Fish and Wildlife at http://ecology.wordpress.com or http://wdfw.wa.gov/ wildlife/management/index.html.

Meet the Species

- Support the pygmy rabbit captive breeding program at the Oregon Zoo by viewing the exhibit or by visiting their Web site at www .oregonzoo.org. You can also watch a video of pygmy rabbits being released into the wild.

COTTON-TOP TAMARIN

Take Action

- Contact Proyecto Tití at www.proyectotiti.com. Through this conservation program you can learn more about cotton-top tamarins, make donations, and purchase eco-products such as mochilas. Purchasing these items will help local communities decrease their dependency on forest products. And with a stable source of income, local communities can help protect the cotton-top tamarin for future generations of Colombians.

CRESTED IBIS

Take Action

- Contact Earth's Endangered Creatures at www.earthsendangered .com to find out more about the crested ibis.

Meet the Species

- Embark on an ecotour to see the crested ibis through Wings Birding Tours Worldwide. Visit http://wingsbirds.com for more information about specific excursions.
- Visit the Yangxian Zhuhuan Nature Preserve. For more information, refer to www.4panda.com/special/bird/site/yangxian.htm.

ECHO PARAKEET

Take Action

- Contact the Parrot Society UK at www.theparrotsocietyuk.org to find out more about echo parakeet conservation efforts and to make donations.
- See "Mauritius Island Birds," below, for more suggestions.

FORMOSAN LANDLOCKED SALMON

Take Action

- Contact Shei-Pa National Park at http://park.org/Taiwan/Government/ Theme/Environmental_Ecological/env64.htm for more information about conservation efforts.

Giant Panda

Take Action

- Contact the Smithsonian National Zoological Park at http://national zoo.si.edu. You can learn more about the species and make donations to the Giant Panda Conservation Fund. This fund will help finance an array of research projects in China and the United States.
- Contact the Conservation and Research for Endangered Species Division of the San Diego Zoo at www.cres.sandiegozoo.org. Their Giant Panda Conservation Unit focuses on the biology and conservation of this species. You can make donations, volunteer, apply for fellowships, and participate in the Adopt the Giant Panda Project.
- Send donations for conservation efforts to:
  The Nature Conservancy
  China Program
  B4-2 Qijiayuan Diplomatic Compound
  No. 9 Jianwai Dajie, Chaoyang District
  Beijing 100600 China
- Donate to research programs devoted to the biology and conservation of giant pandas that are found in four US zoos: San Diego (www.sandiego.org), Atlanta (www.zooatlanta.org), National Zoological Park (www.nationalzoo.si.edu), and Memphis (www.memphiszoo.org). Consult each zoo's Web site to learn more about how to donate and get involved.

Meet the Species

- Embark on an ecotour to see giant pandas and their natural habitats. The Wild Giant Panda Web site at www.wildgiantpanda.com has a listing of nature reserves that offer tourism opportunities. Or visit the Wolong Nature Reserve or Chengdu Panda Breeding Center. For more information about educational excursions, visit www.4panda.com/panda/pandasite/wolong.htm or www.panda.org/cn/english.
- Volunteer at Wolong Nature Preserve in China. Contact Pandas International at www.pandasinternational.org for more information, to make donations, or to sponsor a giant panda.
- Watch a live feed of giant pandas at the Wolong Nature Reserve at www.oiccam.com/webcams/index.html?/panda.
- Visit giant pandas at the National Zoo, Washington, DC, and at the San Diego Zoo. You can check out the new Panda Cam at the San Diego Zoo, which is a live stream from the exhibit.

GOLDEN LION TAMARIN

Take Action

- Learn more about conservation programs at www.micoleao.org.br and www.savethegoldenliontamarin.org. Or visit the National Zoo's Web site, www.natinalzoo.si.edu.

Meet the Species

- Visit the National Zoo or take an eco-tour with Brazil Ecotravel and visit the Poço das Antas Biological Reserve and educational center. For more information, go to the Web site at www.brazil-ecotravel.com.

GRAY WOLF

Take Action

- Contact the Yellowstone Park Foundation at www.ypf.org. Here you can make donations to the general wolf project, sponsor a VHF or GPS collar, or contribute to a community-supported wolf collar.

HAWAIIAN GOOSE (NENE)

Take Action

- Contact the Friends of Hakalau Forest National Wildlife Refuge at www.friendsofhakalauforest.org to make donations to adopt a nene, and to learn about volunteer opportunities to promote natural and cultural conservation efforts in Hawaii.
- Learn more about the nene captive breeding program at Hawaii Volcanoes National Park by visiting www.nps.gov/havo/naturescience/nene.htm. You can also attend an educational field seminar, become a friend of the park by volunteering your time, or make a monetary donation at www.fhvnp.org.
- Drive cautiously in areas marked with NENE CROSSING signs and be mindful of nene on golf courses.
- Observe nene from a distance and never feed them.
- Keep your pets safe at home—they could pose a threat to the nene.

Meet the Species

- Plan a trip to Hakalau Forest National Wildlife Refuge or Hawaii Volcanoes National Park to see the nene and other native birds.
- Visit the nene at the Honolulu Zoo in Hawaii.

IBERIAN LYNX
Take Action

- Contact SOS Lynx at www.soslynx.org to learn more about conservation efforts for this species such as captive breeding programs, and to make donations.
- Support the Cat Specialist Group at www.catsg.org. This organization is sponsored by the World Conservation Union and the Species Survival Commission.
- Visit LifeLince at www.lifelince.org to learn about conservation projects such as radio tracking, camera trapping, and creating supplementary feeding enclosures. There are also volunteer opportunities available.
- Volunteer at an Iberian lynx breeding center (three-month minimum commitment). Visit www.lynxexsitu.es for more information or e-mail lynxexsitu@lynxesitu.es.
- When touring through lynx habitat, especially at Doñana National Park, consider hiking or riding bicycles instead of driving in cars, which have killed a number of lynx in the past.

LORD HOWE'S ISLAND PHASMID
Take Action and Meet the Species

- Visit the Foundation for National Parks and Wildlife at www.fnpw .com.au to learn about Lord Howe's Island phasmid.
- Contact the Lord Howe Island Tourism Association at www.lord howeisland.info to learn about the island's threatened species recovery plan, and to book a trip to see some of the endemic species.

MALA (RUFOUS HARE-WALLABY)
Take Action

- Volunteer to protect the mala and their habitat. Australian Wildlife Conservancy's Scotia Sanctuary offers volunteer opportunities; contact www.australianwildlife.org for more information. Opportunities are also available through the Western Australia Department of Environment at www.dec.wa.gov.au. Or contact the Northern Territory Department of Natural Resources at www.nt.gov.au/nreta/wildlife/programs/volunteers.html for a list of volunteer programs, including ones at Alice Springs Desert Park.
- Conserve water in the desert and remember to share water sources with other animals. That water hole at your camp may be the only water supply that desert animals have access to.

- Be mindful when traveling through the desert. The seeds of grasses that attach to your clothes and car might become invasive weeds elsewhere. Remove them carefully.
- Contact the Marsupial Society of Australia at www.marsupialsociety .org to obtain more information about the threats against mala and other native fauna of Australia.

Meet the Species

- Contact Alice Springs Desert Park at www.alicespringsdesertpark .com.au for more information about this species and to schedule a tour of the park.
- Go to the Shark Bay World Heritage Center in Western Australia. For more information about the mala and its habitat, and to schedule a trip, refer to www.sharkbay.org.

MAURITIUS ISLAND BIRDS: ECHO PARAKEET, MAURITIUS KESTREL, AND PINK PIGEON

Take Action and Meet the Species

- Contact the Mauritian Wildlife Foundation at www.mauritian wildlife.org to learn more about conservation efforts for these species, make donations, volunteer, or obtain information about ecotouring opportunities.
- Contact the African Conservation Foundation at www.african conservation.org to learn more about these species and their habitats, to make donations, or to volunteer for a variety of conservation projects.
- Don't eat seafood. However, if you must, make sure the species is sustainable. The biggest threats to petrels are loss of food resources due to overfishing and the production of fertilizer, the introduction of alien pathogens via fish-farming activities, and the accidental capture of these birds due to legal and illegal fishing techniques. Organizations such as the Environmental Defense Fund at www.edf.org and the Australian Marine Conservation Society at www.amcs.org.au offer sustainable seafood guides.

MILU (PÈRE DAVID'S DEER)

Take Action

- Contact the Whispering Springs Rescue and Research Center at www.whisperingsprings.org. This organization is dedicated to conserving Père David's deer with the hope of reintroducing the herd to the wild. You can see photos of the deer, learn more about the organization's conservation efforts, and make donations to their wish list.

Meet the Species

- Visit Père David's deer at Lake Superior Zoo in Duluth, Minnesota.
- Take a trip to Woburn Abbey in Bedfordshire to see the Père David's deer and other species at this three-thousand-acre park. Go to www .woburnsafari.co.uk.
- Visit Nan Haizi Milu Park (the former Imperial Hunting Park) to see the deer and many other species. For more information, refer to www.beijingjoy.com/attractions/nanhaizimilupark.htm.

NORTHERN BALD IBIS (WALDRAPP)

Take Action

- Be a part of the Waldrappteam Project by visiting www.waldrapp.eu. You can help this bald ibis research and conservation project by joining them for the annual migration, volunteering your time, making donations, or adopting a bird.

PANAMANIAN GOLDEN FROG

Take Action

- Contact the Houston Zoo at www.houstonzoo.org. You can learn more about the golden frog, make donations to the El Valle Amphibian Rescue Center, and watch a video about conservation efforts.
- Support Amphibian Ark—an organization that seeks to ensure the survival of all amphibian species—at www.amphibianark.org. You can learn more about their conservation efforts and make donations.

Meet the Species

- Visit www.ecotourismpanama.com to learn about ecotourism opportunities in Panama. This Web site lists national parks, tour operators, and hotels, and offers a list of endangered species such as the golden frog that tourists may encounter.

PEREGRINE FALCON

Take Action

- Contact the Peregrine Fund: World Center for Birds of Prey at www .peregrinefund.org to learn more about conservation projects, to make donations, to purchase items from the online store, or to contribute your time, skills, and talent to this organization. Volunteer opportunities range from interpretive center docent to bird-sitter to research assistant.

- Visit the Raptor Center of the University of Minnesota at www
.raptor.cvm.umn.edu. You can learn more about falcons, eagles,
hawks, and owls; make donations; become a volunteer; and get information about programs such as Recycling for Raptors.

Meet the Species
- Take a field trip to the Raptor Center located in Saint Paul,
Minnesota.

Pygmy Hog

Take Action
- Contact Durrell Wildlife Conservation Trust at www.durrellwildlife
.org and support their efforts to save this species from extinction.
You can become a member of the organization, make a donation, or
adopt one of the many endangered animals Durrell protects.

Red Wolf

Take Action
- Contact the Red Wolf Coalition at www.redwolves.com to learn
more about the species, to become a member of this organization,
and to make donations to conservation efforts.
- Hunt responsibly. There are one hundred counties in North Carolina, ninety-five of which do not have red wolf populations.
- Drive cautiously since red wolves and other wild animals often cross
roads to find food.
- Keep the roads clean of litter. It attracts wild animals, putting them
in harm's way.

Meet the Species
- Support the Red Wolf Recovery Project sponsored by the US Fish and
Wildlife Service at www.fws.gov/redwolf. You can donate your time as
a volunteer, attend a Howling Safari at Alligator River National Wildlife
Refuge in North Carolina, or attend the Red Wolf Recovery Program
Teacher Workshop held semi-annually to expand red wolf education.
- Visit red wolves at the Point Defiance Zoo and Aquarium in Tacoma,
Washington. You can watch a video about their captive breeding and
recovery program at www.pdza.org.

Short-Tailed Albatross (Steller's Albatross)

Take Action
- Contact the Save the Albatross Campaign at www.savethealbatross
.net. You can support the campaign—sponsored by a number of orga-

nizations such as Bird Life International—by educating yourself about the challenges facing conservationists, by signing up for an e-newsletter, by participating in the used-stamp project whose proceeds directly benefit the albatross, and by making donations.

- Make sure you properly discard garbage, particularly plastic and balloons. Albatross and other seabirds can become entangled in the strings, while other animals may ingest them and die.
- Don't discard used oil into city sewers or municipal water supplies, as it can end up in the ocean. If the bird's feathers become oiled, it can be lethal, because they will no longer be waterproof.

SUMATRAN RHINO

Take Action

- Visit the International Rhino Foundation at www.rhinos-irf.org to learn more about what you can do to help save this endangered species, to sign up for a newsletter, to make donations, and to adopt a rhino.
- Support the Center for Conservation and Research of Endangered Wildlife at the Cincinnati Zoo (CREW). The organization sponsors Sumatran rhino captive breeding and cooperates with Indonesian conservationists to protect the rhino's remaining forest habitat. CREW also supports Rhino Protection Units on Sumatra, Borneo, and on the Malayan Peninsula. Learn more at www.cincinnatizoo .org/conservation.
- Don't purchase animal products such as rhino horn. One of the biggest threats to Sumatran rhinos is illegal poaching.

Meet the Species

- Visit the Sumatran rhino exhibit at the Cincinnati Zoo or watch them via rhinocam on the zoo Web site at www.cincinnatizoo.org. In addition to sponsoring a groundbreaking captive breeding program, the Carl. H. Lindner Jr. Family Center for Conservation and Research of Endangered Wildlife (CREW) at the zoo also supports Rhino Protection Units to save them from poachers. There are opportunities for you to get involved through their volunteer program.

TAHKI (PRZEWALSKI'S HORSE)

Take Action

- Contact the Foundation for the Preservation and Protection of the Przewalski Horse at www.treemail.nl/takh to learn more about this species and its habitat.

Meet the Species

- Visit Hustai National Park in Mongolia to see the Przewalski's horse. Go to www.hustai.mn to read details about tours, accommodations, and eco-volunteer or research opportunities.

VANCOUVER ISLAND MARMOT

Take Action

- Contact the Marmot Recovery Foundation at www.marmots.org to learn more about what you can do to help save this species from extinction, to participate in their observer program, to make a donation, and to adopt a marmot for yourself or a loved one.

WHOOPING CRANE

Take Action

- Support the International Crane Foundation at www.savingcranes .org by learning more about crane conservation, making donations, joining the organization, becoming a volunteer, or visiting the foundation, located in Baraboo, Wisconsin. Guided tours are available.
- Contact Operation Migration at www.operationmigration.org. This organization developed the technique of human-led migration and is responsible for the care and training of each new generation of reintroduced whooping cranes from the time they hatch until they are released in Florida. Visit this Web site to learn more about Operation Migration's reintroduction programs, to make donations, to become a volunteer, to watch video clips, and to receive daily updates written by the pilots and ground crew who are dedicated to protecting the whooping crane and other endangered migratory birds.

Meet the Species

- Participate in the Annual Midwest Crane Count in the spring. Visit www.savingcranes.org for more information. You can see sandhill cranes and whooping cranes while contributing to conservation efforts.
- Visit the Aransas Wildlife Refuge in Austwell, Texas, to see the largest wild flock of whooping cranes in the winter. For more information, contact the US Fish and Wildlife Service at www.fws.gov/ southwest/REFUGES/texas/aransas.

YARIGUIES BRUSH FINCH

Take Action

- Volunteer with ProAves, Colombia's leading bird conservation organization. Volunteer opportunities include banding birds or installing and monitoring artificial nests. Or make donations to ProAves to purchase and protect forest land where endangered species, such as the Yariguies brush finch, dwell. For more information, visit the English-language Web site at www.proaves.org.

- Support the Natural History Museum, UK, host to one of the world's largest biological collections. Researchers, students, and visitors from around the world have used the museum's collections to understand the world's biodiversity. For information about volunteering or donations visit www.nhm.ac.uk.

Meet the Species

- Go birding in Colombia through ProAve's partner Eco Turs. Visit the Web site at www.ecoturs.com, contact them directly by e-mail at info@ecoturs.org, or call the US phone number: 540-878-5410.

Zino's Petrel

Take Action

- Avoid littering at sea, which kills many seabirds, including Zino's petrel. For instance, polystyrene (like you find in Styrofoam food packaging) breaks up into small white balls, which seabirds swallow, causing intestinal blocks and death.
- Hike respectfully. Never leave trash or bits of food on the trails or wilderness areas of Madeira, as this encourages rats and cats, who are predators of petrels.

Meet the Species

- Take a nighttime guided tour to Pico Areeiro on Madeira, where you can hear the Zino's petrel and maybe see one flashing by on a moonlit night. Contact Madeira Wind Birds at www.madeirabirds.com or Ventura do Mar at www.venturadomar.com for information.

PLANTS

For many of the endangered plant species in this book, you will find the following resources helpful in obtaining more information about conservation activities.

- The Royal Botanic Gardens, Kew, is a leader in the conservation of endangered plants and their habitats, and is directly involved with many of the species mentioned in this book. For more information, to make donations, to plan a trip to the gardens, and to become a friend of the Kew, please visit www.kew.org/conservation. Also visit Wakehurst Place—Kew's three-hundred-acre country garden and home to the Millennium Seed Bank at www.nationaltrust.org .uk/main/w-wakehurstplace.

- The Center for Plant Conservation strives to conserve and restore native plant species of the United States and is currently involved in conservation efforts with many of the plants discussed in *Hope for Animals*. For more information, to make donations, and to join the plant sponsorship program, visit www.centerforplantconservation.org.

CAFÉ MARRON

Take Action

- Contact the Mauritian Wildlife Foundation at www.mauritian wildlife.org to make donations and learn more about this organization's conservation projects on Rodrigues Island.

TAHINA PALM

Take Action

- To learn more about the Tahina palm and to view images, visit www .rarepalmseeds.com
- Contact Madagascar Wildlife Conservation at www.mwc-info.net/en to learn more and to make donations. This organization involves local communities in all its conservation projects.

WOLLEMI PINE

Take Action and Meet the Species

- Contact Wollemi Australia at www.wollemipine.com to learn more about this species, to join their conservation club, and to order your own Wollemi pine. An important conservation strategy to protect wild populations of these plants is to ensure that Wollemi pines are growing in other areas around the globe—whether that's a national park or in private collections.
- Take a trip to Wollemi National Park located north of the Blue Mountains in Australia. Or visit the United States Botanic Garden in Washington, D.C.

HABITATS

GOMBE: TANZANIA, AFRICA

Take Action

- Support the Jane Goodall Institute's TACARE Program. Visit JGI's Web site and look up the Institute's Africa programs to discover how you can support restoration of habitat in and around Gombe.

Guadalupe Island: Baja California, Mexico

Take Action

- Support the work of Grupo de Ecología y Conservacíon de Islas, a Mexican nonprofit devoted to the restoration of Mexican islands, by making donations or becoming a volunteer. You can contact Dr. Alfonso Aguirre Munoz, the director general, at alfonso.arguirre@conservaciondeislas.org; or write to:

    Grupo de Ecología y Conservacíon de Islas, AC
    Avenida Lopez Mateos 1590–3
    Fracc. Playa Ensenada
    Ensenada, Baja California 22880

- Contact Nadia Olivares, director of the biosphere reserve on the island, at islaguadalupe@conanp.gob.mx for more information about how you can help.
- Learn more by contacting Island Conservation at www.island conservation.org. This nonprofit organization is dedicated to protecting island ecosystems and is one of the groups currently working to preserve Guadalupe Island. You can make general donations or become a volunteer.
- Wear clean shoes and clothes when visiting islands in order to prevent exotic species invasion.
- Do not take pets to the island and ensure, if you travel by boat, that your vessel is rat-free.

Ecotourism

- Travel to Guadalupe Island through Horizon Charters—a company committed to sustainable ecotourism. For more information about specific excursions, go to www.horizoncharters.com.

Kenya Coast, Africa

Take Action

- Contact the Baobab Trust at www.thebaobabtrust.com to learn more about wildlife and habitat conservation.
- Support the Haller Foundation by visiting www.thehallerfoundation .com. This organization strives to create sustainable, ecologically sound communities in Africa. You can learn more about their conservation projects or make donations.
- Learn more about land reclamation in Haller Park by contacting Bamburi Cement at www.bamburicement.com.
- Get involved with the Green Belt Movement by visiting www.green beltmovement.org. Though this organization's home base is in Kenya,

they have a very active international program as well. The Green Belt Movement focuses on environmental conservation and tree planting among other green community-building projects. Donations are welcome.

LOESS PLATEAU: NORTHWEST PROVINCES, CHINA

Take Action

- Learn more about the Watershed Rehabilitation Project at the Loess Plateau by contacting the World Bank at http://web.worldbank.org. There is a slide show and educational video available as well.
- Contact the EroChina Soil Erosion Project at www.erochina.alterra .nl to learn more about what's being done to stop soil erosion on the plateau.

SUDBURY, ONTARIO

Take Action

- Contact the city of Greater Sudbury at www.greatersudbury.ca to learn more about their Regreening Program and to find out how you can get involved.

Index

Page numbers of photos appear in italics.

Reading Group Guide

Discussion Questions

Considering its warm, conversational tone and its depth of timely information, *Hope for Animals and Their World* makes an excellent book for book groups to read and discuss. The following fifteen questions will help kick-start your discussion.

1. Of the many success stories profiled in this book, which ones inspired you the most?

2. When the California condor became seriously endangered the "protectionist" wanted to give the birds better protection in the wild and if this did not work, let them gradually disappear with dignity. "Interventionists" wanted to save the birds through captive breeding with the hope that they could be successfully re-released back into the wild. Given what you know of the captive-bred condors' challenges re-acclimating to life in the wild, where do you stand on this debate?

3. Some biologists feel strongly that the animals they study should be identified by numbers, believing this practice is more scientific. For instance, the whooping cranes in the Operation Migration are known only by numbers. Some biologists also argue that giving animals numbers instead of names creates emotional distance for the scientists as well as the general public—helping everyone handle the inevitable losses of individuals in captivity and in the wild. Other biologists feel equally strongly that animals should be identified by names. On which side of this debate do you stand?

4. If an animal species no longer has a suitable habitat on Earth to sustain it, do you think it should still be saved through captive breeding? Why or why not?

5. In her introduction to Part 2, Jane describes the reaction many people have had to her efforts to protect all of Earth's creatures, not just the "charismatic ones," as she puts it. She recalls being asked: "Why on Earth would anyone devote themselves to protecting a bug?" Do you think saving different species requires prioritization?

6. Do you think it's ethically correct to eradicate cats, rats, and other invasive species from our islands in order to save endemic species from extinction?

7. In Part 2, Jane describes feeling "an almost spiritual level of connectedness" to the whooping crane when she is flying in the ultra-light aircraft. Does Jane have other moments of spiritual connection in this book? Do you think science and spirituality are linked in any way? Why or why not?

8. Until you read the chapter "Healing Earth's Scars" were you aware of how successfully we can restore a ravaged habitat? Have you seen other examples of this kind of land and water restoration in your life?

9. Many readers were amazed to learn that some animals and plants that we assumed were extinct are now being re-discovered in remote areas of the planet, and that completely new species of animals and plants are still being discovered on Earth. Did this surprise you? What else in the book surprised you?

10. On page 225, Jane describes a conflict between "concern for the individual and concern for the future of a species." How do you understand this conflict? On what side of this debate does Jane fall? Do you agree with her reasoning?

11. Throughout the book Jane references young people's innate attraction to the natural world, including her own girlhood fascination with animals and plants. What were the plants, trees, or animals that

you were drawn to as a youth? Did those interests shape your adult relationship to the natural world? If so, how?

12. In the chapter "Why Save Endangered Species?" Jane writes about being motivated by her love of animals, such as the infant chimpanzee Flint, and how other biologists are motivated by their love for the endangered animals they protect. Is there an animal or plant that you feel this kind of love for? What lengths would you go to in order to protect it from extinction?

13. Have you ever seen a critically endangered animal or plant in the natural world? In captivity? What was that experience like?

14. At the very end of the book, Jane outlines her four reasons for hope. Do you agree with her? Are there other reasons you would add? Do you feel more optimistic about the state of our planet after reading this book?

15. How does this book inspire you to take action in your everyday life? What will you do differently now that you've read this book?